**FIRST**
**Robots**
Behind the Design

ROCKPORT

First published in the United States of America by
Rockport Publishers, a member of
Quayside Publishing Group
33 Commercial Street
Gloucester, Massachusetts 01930-5089
Telephone: (978) 282-9590
Fax: (978) 283-2742
www.rockpub.com

**Library of Congress Cataloging-in-Publication Data available**

ISBN-13: 978-1-59253-366-4
ISBN-10: 1-59253-366-3

10 9 8 7 6 5 4 3 2 1

Cover Design & Design: Collaborated, Inc.
Cover Image: FIRST award-winning teams, including Team 357

Printed in USA

# FIRST
# Robots
## Behind the Design

ROCKPORT

PUBLISHERS

➡ Vince Wilczynski

➡ Stephanie Slezycki

➡ Foreword by Dean Kamen

➡ Afterword by Woodie Flowers

# CONTENTS

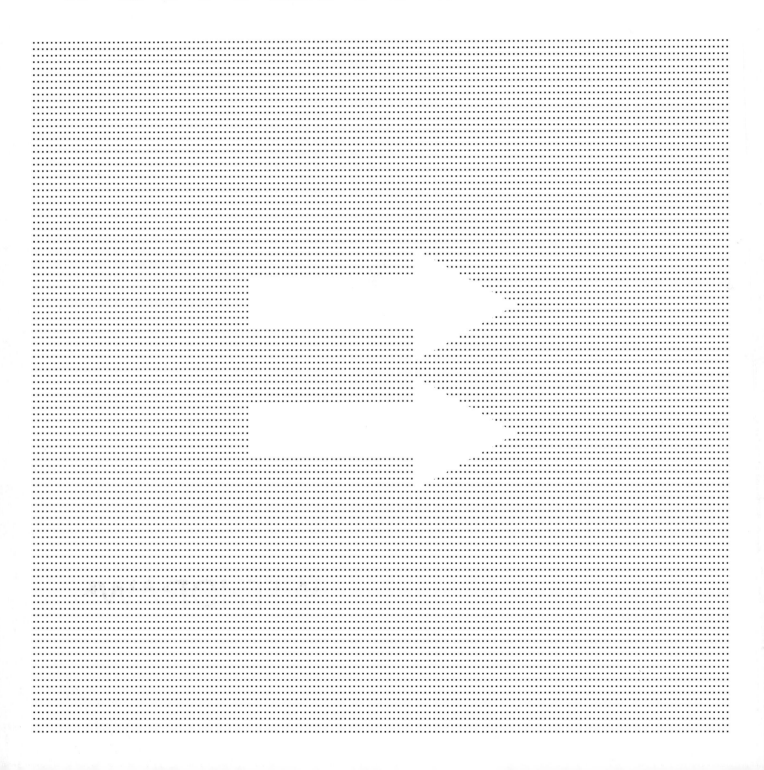

# FOREWORD

**by Dean Kamen**

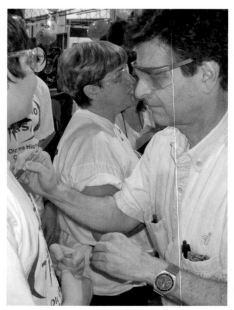

⇄ ⇈ **FIRST Founder Dean Kamen is relentless in his pursuit to transform attitudes about science and technology. FIRST Robotics Competitions celebrate engineering, promote personal development, and foster friendship among participants. Showing their appreciation of his work, students and mentors collect Dean's autograph, which he willingly provides by signing their team uniforms and competition medals.**

# ROBOTICS: INNOVATION AND INSPIRATION

Designing solutions to problems is great fun. Problem solving has helped our species evolve from being cave dwellers to space explorers, advancing society along the way by continually raising our standard of living.

The only laws that apply in such pursuits are the laws of nature, and all else is optimization—a trade-off of the factors associated with possible solutions. For any problem, there are an infinite number of solutions, with each solution resulting from a finite set of resources. Finding the balance between the finite and infinite is what makes design rewarding.

Breakthroughs in thinking provide new perspectives and lead to innovation. But even the most creative solutions are bound by realities that must be applied before innovation is realized. For many products, time, cost, and materials are boundary conditions which must be honored to progress from an idea to invention.

Fifteen years ago, we saw a problem: Not enough young people were recognizing the opportunity and joy that comes with a lifetime of solving problems through science, technology, and engineering. That's when we founded FIRST and created the FIRST Robotics Competition.

The FIRST Robotics Competition inspires high school students to directly experience science and technology, fostering creativity and discovery through hands-on exercises to solve real-world problems. We provide a problem to solve, and students work like crazy with mentors to design and build their robotic solutions.

Borrowing a page from the sports playbook, we design intriguing competitions where robots compete with and against one another. Teams are formed to solve the design challenge and build robots for the competition. In the process, students are exposed to some of today's best role models: engineers and technologists who improve society, and, along the way, inspire students.

Unlike sports, however, our purpose is not to see which team can accumulate the most points on a competition field. Instead, we are dedicated to changing the way these students think about their world. We strive to instill an understanding that cooperation can accompany competition. We work to transform the culture into one that recognizes the value and importance of science and technology in our advanced world.

While FIRST teams immensely enjoy the on-field competition with their robotic creations, they covet the chance to be recognized as the team that best achieves FIRST's goal to transform its community into one which values science and technology and inspires students to pursue technical careers. To that end, each competition recognizes outstanding performance in creativity, entrepreneurship, spirit, design, and other areas. At our 33 Regional Competitions and the Championship, teams are evaluated by a panel of judges who closely review the robots and teams to determine the best for each award category.

In 2006, more than 1,100 teams built robots for the competition. This book profiles 30 of the finest robots and the teams responsible for them. Each team was judged most impressive in the areas of innovative control, quality, creativity, advanced technology, or industrial design. For each award category, six of the 34 winning teams from the regional and championship competitions are detailed in this book—further distinction for their terrific work.

FIRST demonstrates the reality that young people respond to tough challenges not with excuses and distractions but with energy and innovative thinking. We celebrate the simple truth that in six weeks an adult engineer can provide a young person with learning and inspiration that can last a lifetime. The FIRST experience helps students discover the rewards and excitement of science and technology. As a result, FIRST students don't avoid education in science and technology; they seek it.

I find their work inspiring, and hope you do as well.

*Dean Kamen is President of DEKA Research & Development Corporation, a dynamic company focused on the development of revolutionary new technologies that span a diverse set of applications. As an inventor, physicist, and entrepreneur, Kamen has dedicated his life to developing technologies that help people lead better lives. One of Kamen's proudest accomplishments is founding FIRST.*

# INTRODUCTION

The world of robotics is rapidly expanding and already spans diverse fields such as medicine, education, entertainment, automation, and transportation. Robotic applications are becoming common; no longer are robots mere fantasy in science fiction, but we find them in our homes, schools, and businesses.

Robots are integrated engineering systems: a combination of mechanical, electrical, electronic, and computer technologies. They have been used to inspire students and are popular tools for learning for several reasons: availability of components to build them, our familiarity with their use, and the multidisciplinary nature of their design.

The FIRST Robotics Competition is one example of using robots to inspire students. FIRST is an acronym: For Inspiration of Science and Technology. The premise of FIRST is to partner youth with practicing professionals to solve challenging engineering problems. FIRST Robotics Competition participants build sophisticated robotic devices that compete in mechanical sports.

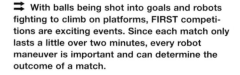 With balls being shot into goals and robots fighting to climb on platforms, FIRST competitions are exciting events. Since each match only lasts a little over two minutes, every robot maneuver is important and can determine the outcome of a match.

➡ Three teams form an alliance for each match, where they compete against an opposing alliance of teams. Each alliance is random; an alliance member in one match could be an opponent in the next match.

FIRST teams consist of high school students and faculty, engineers, technicians, business leaders, university students and staff, and concerned citizens. Each year, a new "FIRST game" is developed which requires robots to perform a variety of tasks such as moving across a field, climbing ramps, hanging from bars, and placing objects in goals. In addition to being manually operated by remote control, the robots also operate under autonomous control for a segment of each match.

Teams only have six weeks from when the annual game is announced and they receive the Kit of Parts to design, build, program, and test their robots. Following these six weeks, teams compete at regional competitions held in 32 locations in North America and one in Israel. The competitions are wildly exciting, energetic, and motivating.

FIRST competitions are unique in their structure and purpose. Alliances of robots compete in each match, so teams have to work together to play the game. Competition on the field is paired with cooperation between teams to create an atmosphere of gracious professionalism. The FIRST Robotics Competition Awards support this goal;

most are based on attributes such as spirit, entrepreneurship, inspiration, teamwork, and effectiveness. The teams compete aggressively but treat each other kindly in the process.

The judging panels at each competition included engineers, educators, technologists, and business and government leaders. The judges' review is comprehensive, and they visit with each team and observe the robots in competition.

This book salutes the high levels of creativity, ingenuity, and design that go into building FIRST robots. The 30 robot teams detailed in this book are examples of superior design in five categories: control, quality, creativity, advanced technology, and industrial design. Winning an award at a FIRST Robotics Competition is a great accomplishment and an acknowledgement of superior performance by the team. A list of participants from each profiled team is included in this book.

As exemplified by their willingness to share their design successes, these teams are fulfilling FIRST's mission to inspire others. We hope the assembled work serves as a template for learning about robot design, and that it is a resource for all who are interested in the field of robotics.

⬆⬆ The playing field in front of the goals is a popular place for robots. Half of the robots want to be in front of the goal to score, but by doing so they draw the attention of defending robots.

⇄ Offense and defense are both important parts of the game. Defending robots maneuver to stop the most pressing scoring threats from opposing robots that are shooting or climbing the ramp.

⇄ The driver's station is a safe place to be for human competitors, protecting them from robot collisions and the occasional flurry of flying robot parts.

# 2006 FIRST ROBOTICS COMPETITION: AIM HIGH

## PUSHING TEAMS TO NEW HEIGHTS

The 2006 FIRST Robotics Competition challenged robots to compete in a game that could be described as a combination of robot-basketball, robot-soccer, and robot-king-of-the-hill. For each match, three robots formed an alliance to shoot balls into a raised goal, deposit balls in a low goal, climb a ramp onto a platform, and prevent the opposing alliance of three robots from accomplishing the same tasks.

### ⇉ The Playing Field and Goals

Played on a 54-foot (16.4-m)-by-26-foot (7.9-m) field with six robots competing during each match, the on-field action was nonstop. Three goals were available at each end of the field, with one set of goals designated for each alliance. Two lower goals were located in the corners of the field for robots to push or shoot balls into. The lower goal opening was a small distance off the playing field floor, with an incline in front of the goal as an obstacle. The upper goal was a 2.5-foot (0.8-m)-diameter target, centered 8.5 feet (2.6 m) above the playing surface and mounted above the opposing alliance's driver station. A foot-high ramp and raised platform—wide enough for only three robots to climb on at any one time—provided a final game challenge.

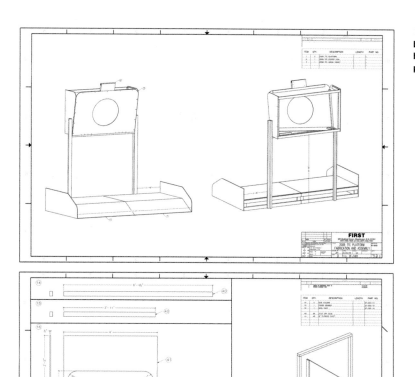

⇄ Detailed blueprints provide instructions on building the game playing field. In addition to building a robot, teams often construct a full-size playing field to test their concepts on.

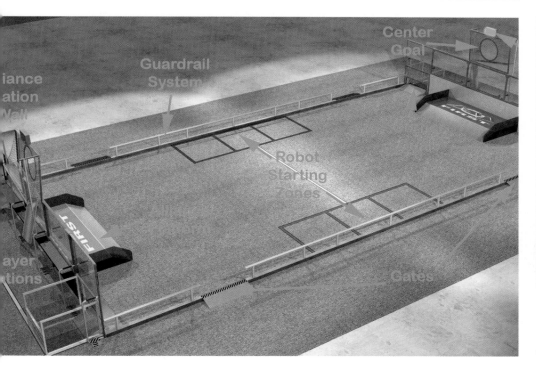

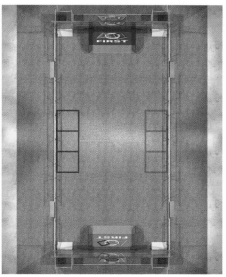

⇄ ⇈ A large, open playing field leaves plenty of room for robots. Each alliance starts in a designated position and races to their goals at the field ends to score points.

## ⇄ Scoring Points

Teams earned points by shooting or pushing 8-inch (20.3 cm)-diameter foam balls into the goals. A ball in the upper goal was worth three points, while each ball in a lower goal was worth one point. The ramp and platform beneath the upper goal counted only at the end of the match, when a single robot on the ramp earned the alliance five points; two robots were worth 10 points; and three robots on the structure earned the alliance 25 points.

Robots were restricted to a height of 5 feet (1.5 m), thereby requiring them to shoot—not drop—balls into the higher goals. This size restriction also prevented robots from blocking the opposing alliance's upper goal.

## ⇄ Match Play: Alliances of Three Robots

The six robots competing in each match were divided into two alliances, with three teams per alliance. An alliance would work together for the match duration to develop and execute a scoring strategy. Alliances were randomly assigned for each match, which helped foster cooperation in the midst of competition.

Two robot operators for each team were located at one end of the field, with their goals at the other end. Skilled driving was needed to score points in goals located more than 50 feet (15.2 m) away. In addition to the robot operators, each team also included a coach and a human player. It was the responsibility of the three human players on each alliance to throw balls to the robots. The coaches kept watch on the play and advised the robot operators where best to position each team's robot.

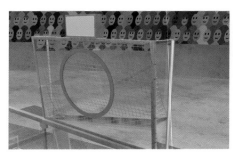

⇡⇡ Located well off the playing field floor, the upper goal is a big target. The green light, a favorite color for the on-board cameras, indicates the goal is open for the allianceto score.

⇡⇡ The lower goals are an easier target, but the balls scored here are only worth one-third the value of the balls scored in the upper goals. The ramp and platform are an enticing destination for robots at the end of the match to earn additional points.

⇄ Robots scored by getting balls into an upper goal illuminated by a target light, or into the lower goals located in the corners of the field next to the ramps. Teams of three robots form the Red Alliance to play three teams in the Blue Alliance, with each alliance assigned its own goals.

⇡⇡ The coin of the realm for the competition is foam balls. Robots start with 10 balls and can reload by gathering balls from either the field or from their human players. The balls must be propelled into the upper goal or rolled into the lower goal.

⇡⇣ A 30-degree angle stands between the playing field floor and the platform. Robots not designed with sufficient stability guarantee plenty of excitement while climbing this ramp.

⇡⇡ A small ramp provides a small obstacle in front of the lower goals. Balls must roll over the ramp and into the goal to score.

## ⇄ Four Periods per Match

Play began with the robots operating autonomously, followed by periods of remote control. Each match was divided into four periods to encourage multiple strategies of play: the initial autonomous period when both alliances could score, a second and third period when only one alliance was allowed to score, and the final period when both alliances could score. The autonomous period was ten seconds, and each of the remaining periods lasted 40 seconds.

For the initial autonomous period, each robot started with ten balls, and their controllers were programmed with maneuvers to score the balls or play defense. A green light above the upper goals indicated when they were open for scoring. The light also served as a target for cameras mounted on the robots to find the goals. Well-programmed robots used their cameras to aim and fire at the target automatically.

During the second and third periods, only one alliance's goals were open for scoring. One alliance was designated on offense, with the other on defense.

The green light above the upper goal was illuminated to indicate when that goal was active.

While playing defense, teams were not allowed to score. To give the offense an advantage during the second and third periods, only two robots from the defensive alliance could play defense, with the remaining robot relegated to the other half of the field. While in the far end of the field, the third defensive robot would often prepare for the next period of scoring.

During the final period, all goals were open and there were no restrictions on where robots could operate. The high point values for robots on the platforms at the end of the match tempted teams to abandon scoring in their goals and head across the field to their home platforms. This scoring option was a natural cliff-hanger for action, as alliances often dashed the full length of the field in the final seconds of a match to climb their home platforms and score up to 25 points.

## ⇄ A Winning Strategy

Two key advantages were obtained by the alliance that scored the most points during the autonomous period. First, that alliance was awarded an additional ten points. Second, that alliance played defense first, thereby requiring one of their robots to stay in the far end of the field. In the backfield, this robot could collect balls and prepare for the next period when its alliance would be allowed to score points. When the goals opened at the start of the third period, they would remain open for the rest of the match, thereby allowing the alliance that won the autonomous period to be in scoring positions for two periods in a row.

The game rules and field structure required robust and clever design. Effective robots had to be capable of multiple tasks, and skilled driving was needed to collect balls, shoot balls, defend against other robots, and climb on the platform—all in 2 minutes and ten seconds. And those tasks had to be completed while three other robots were doing their best to prevent their opponents being successful.

⇈ Ten seconds of autonomous play start each match. Robots are programmed to score in the upper and lower goals during this period, but they must be careful to avoid defensively programmed robots that challenge their scoring opportunity.

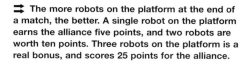

⇒ The more robots on the platform at the end of a match, the better. A single robot on the platform earns the alliance five points, and two robots are worth ten points. Three robots on the platform is a real bonus, and scores 25 points for the alliance.

⇈ Climbing the ramp onto a platform earns additional points at the end of each match.

# THE FIRST KIT OF PARTS

## A STARTING POINT FOR ROBOT DESIGN

Creating a level playing field is an important aspect of the FIRST Robotics Competition. To help achieve that goal, a FIRST Kit of Parts is supplied to all teams at the beginning of the competition season. This kit contains the raw materials to build a basic robot, including structural framing, motors, pneumatics, electronics, and control components. Teams may augment this list of supplies with additional materials, such as aluminum stock and polycarbonate, to construct their robots.

The FIRST Kit contains more than 125 individual components, with multiple quantities of most kit components. These items range from all the parts needed to construct a robot chassis to the electrical connectors that fasten wires together. Using the kit and a FIRST-written manual titled *Robot Construction Tips and Tricks*, FIRST participants learn how to assemble and integrate the mechanical, electrical, pneumatic, and control components of a robot.

The FIRST Kit is assembled each year from a variety of sources. Some material in the kit is donated by corporations that share FIRST's vision to make engineering accessible and attractive to today's youth. The material includes hardware, such as motors and electronics, and software, such as National Instruments' LAB-View data-acquisition software and Autodesk's Inventor and 3-D Studio-Max programs for mechanical design and computer animation. Other material deemed integral to the game, such as a digital camera for target-image acquisition and processing, is procured by FIRST and added to the package. The FIRST Kit changes each year based on the donated supplies and needs of that year's game.

⬆⬆ The FIRST Kit of Parts is distributed in bins and boxes to all teams on the same day the game is announced. The teams then have a little over six weeks to transform the contents of these bins into a working robot.

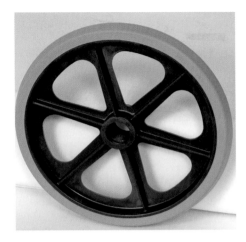

⇄ The kit contains a wide variety of materials, including electronics, aluminum framing, fasteners, motors, transmissions, a pneumatic system, and even wheels.

↓↓ While there are no instructions for building a robot in the kit of parts, instructions are provided to assist with the assembly of components and to provide wiring specifications.

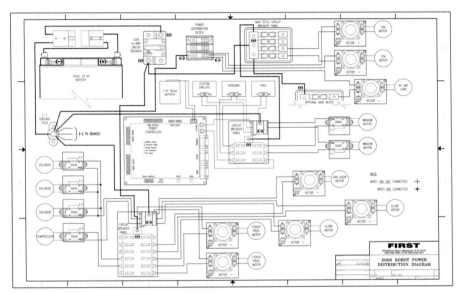

↑↑ Students sort through the bins to unpack components, take inventory of the raw materials, and begin to think about how the parts can be assembled to build robotic mechanisms.

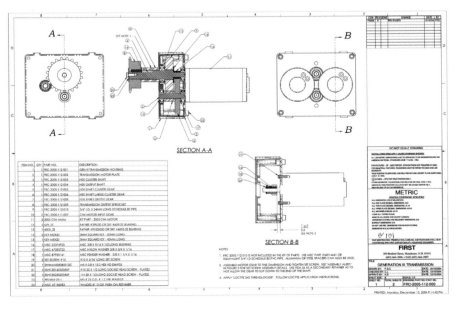

# ⇄ Robot Chassis System

The robot chassis, also referred to as the robot base, is the foundation of the machine. The chassis serves as a frame to which all components of the robot attach, including the drive wheels, propulsion motors, and all robot appendages. The chassis system consists of structural aluminum members that can be configured in a number of different ways to form a solid robot base. This set of components was designed by FIRST robot enthusiasts and is exclusively available to FIRST Robotics Competition teams.

The largest components of the chassis system are 38-inch (96.5-cm) aluminum frame pieces that are used as the sides of the robot base.

These pieces are predrilled with holes spaced every inch (2.54 cm) to allow the frame to be assembled in any number of configurations, or for mounting additional hardware on the robot frame.

⬇⬇ The kit includes components to construct a robot chassis. The frame can be constructed in a variety of configurations to meet the team's design requirements.

Since the holes are drilled in each face of the pieces, any side of the frame can be a mounting surface. Individual pieces are joined together with standard screws and bolts that are supplied in the kit.

The chassis system is strong and versatile. Usually, two structural members are used on either side of the robot to support the robot's wheels and provide overall strength. The chassis system includes end cap pieces to join the side structural members and create a box frame where the wheels can be mounted. The two wheel modules are connected with front and rear bumpers (also made of structural aluminum) to form the robot frame.

Hardware is included to mount the FIRST-supplied drive system transmissions to the chassis. An aluminum plate connects the upper surface of each transmission and adds rigidity to the robot frame. The resulting frame is extremely durable and has proven to be well designed for FIRST Robotics Competitions.

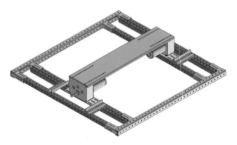

⬆⬆ Detailed assembly instructions help teams learn how to build a strong chassis that can withstand the demands of a FIRST competition.

⬆⬆ Sufficient materials are included in the kit to allow teams to construct a functional, unique robot base.

# ⇄ Robot Drive Train

The robot drive train converts the power from the drive motors to motion in the robot drive system. The parts to assemble a complete robot drive train, including motors, transmission, sprockets, shafts, chain, and couplings, are included in the FIRST Kit. These parts can be assembled in a variety of ways, depending on the team's propulsion requirement.

Transferring the power from the drive motors to the wheels generally requires a transmission to convert the high-speed, low-torque output of the motor to a lower-speed, higher-torque output that can be transmitted to the drive wheels. The 2006 FIRST Kit contained a pair of transmission kits designed to reduce the speeds of the largest motors in the FIRST Kit to typical speeds for a drive system. The robot transmission supplied in the 2006 FIRST Kit of Parts reduced motor output speed by a factor of 13, and thereby increased the output torque by a factor of 13.

The 2006 FIRST transmissions were a common component on many robots because of the high-quality and ease of integration on FIRST robots. However teams were not required to use the transmissions supplied in the FIRST Kit of Parts. Instead, some teams designed and constructed their own transmissions, while others purchased commercial transmissions for their robots.

⬆⬆ Components to build two transmissions are included in each FIRST Kit. The transmissions were specifically designed for FIRST teams.

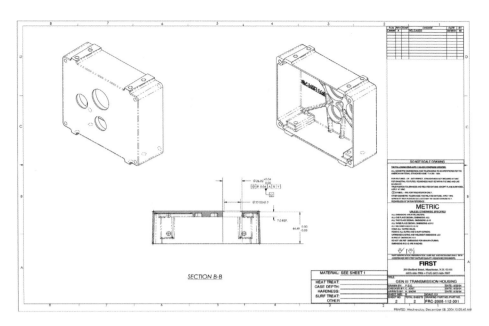

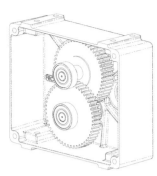

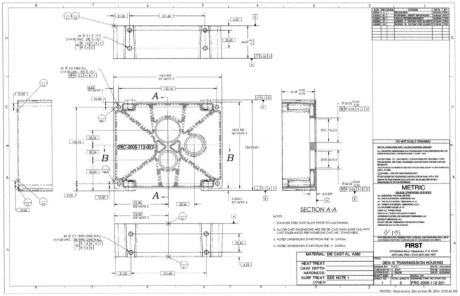

**≡ ↑↑** Two transmissions are included in the kit to reduce motor speed and increase the output torque. An instruction manual details the many assembly steps to convert the gears, spacers, shafts, and mounting box into a working transmission.

**↑↑** The parts are assembled to build each transmission. With dozens of parts, it is important to use care during assembly.

**↑↑** The transmissions are commonly used as part of the robot drive system. Two motors are often used for each side of a robot's drive system.

# ⇄ FIRST Robotics Competition Motors

The motors supplied in the FIRST Kit are the only motors allowed on FIRST robots. This rule encourages innovation and ensures that all robots have the same maximum power capacity.

The kit's direct current motors (powered from a 12-volt battery supplied in the kit) range in power from 17 to 300 watts. Each motor has a commercial application in products such as mobile-home trailer jacks, van door closing systems, car windows, gear-switching mechanisms in automobile transmissions, and ride-on vehicles for children.

Performance specifications for motor speed, torque, weight, and physical size are detailed in the documentation supplied to each team. Teams evaluate their robot functions and select the best motor for each purpose. The motors cannot be altered, and teams must design the motor couplings, gear reductions, and transmissions to use each motor.

⇈ A collection of motors are included in the FIRST Kit. Some include an attached transmission, but others just consist of the motor and an output shaft. Determining how to use the motors to achieve robot functions is a task faced by every team in the competition.

# ⇉ Robot Power and Control Systems

The robot power-distribution system provided in the kit ensures that power is safely supplied to each motor. The 12-volt batteries in the kit are normally used on motorcycles and are commercially available. A 120-ampere main circuit breaker protects the overall system, and individual circuit breakers ranging in size from 20-40 amperes protect each motor. Wiring—and the connections between each motor and the robot distribution panel—must meet industrial standards for each application.

Voltage controllers are supplied in the kit to regulate the power delivered to each motor. The voltage controllers supply a specified voltage level to each motor and enable each motor to operate at any speed. Voltage relays, which provide full power in either the forward or reverse directions, are also included in the FIRST Kit for those applications where proportional speed control is not needed.

The Innovation First robot control system is another important component supplied in the FIRST Kit. There are two principal parts of this system: the operator interface and the robot controller. The operator interface converts input from the driver-controlled joysticks, and control buttons into signals that can be transmitted to the robot. The operator interface also displays feedback from the robot, such as the battery power and the sensor output to enable the robot drivers to monitor these conditions while driving the robot.

The robot controller acts as the brain, and directs all actions of the robot. Commands sent from the robot drivers are transmitted through a wireless radio and received by the robot controller. The controller converts these incoming commands to electrical signals delivered to the voltage relays and con-trollers. The robot controller also receives input from on-board sensors, and interprets these signals to direct robot functions.

The robot controller's output is determined by the controller program. The programming instructions are written in the "C" programming language and can be modified by teams to meet their requirements. The controller can also be programmed to operate the robot autonomously when input from on-board sensors determines robot functions.

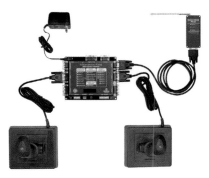

⇈ All motors are powered by either a "Spike" voltage relay or a "Victor" speed controller. These units interpret signals provided by the robot controller to deliver either a constant or a variable voltage to each motor on the robot.

⇈ Robot operators use joysticks, dials, and switches connected to the operator interface to command robot actions. A radio transmits these signals to the robot.

⇈ A peek under the hood shows the sophisticated collection of electronics in a voltage controller. The relays and controllers are specially designed and manufactured for FIRST teams.

⇈ The robot controller is connected to a radio to receive transmitted commands from the robot operators. A small backup battery powers the unit if the robot's main battery is low.

⇄ The Innovation First operator interface and robot controller serve as the robot brain. These units receive signals from the drivers and on-board sensors, and interpret the gathered information to direct robot functions.

## ⇄ Robot Pneumatics

A complete pneumatics system, including an air compressor, storage cylinders, solenoids, pistons, and monitoring sensors are provided in the FIRST Kit. This system allows teams to convert electrical energy into pneumatic energy by compressing and storing air. Solenoids (electronically actuated valves) release the stored air when prompted by the robot controller to move the pistons. Pneumatic force is commonly used for the linear motion of a robot appendage.

## ⇄ Robot Sensors

The FIRST Kit includes a variety of sensors that can be used to measure conditions on the robot. The sensors are connected to the robot controller, which determines how the robot should act based on sensor readings.

The sensors either measure the relative value of a parameter, or function like a switch. Potentiometers (to measure distance), accelerometers (to measure acceleration), and gyro rate sensors (to measure angular motion) are sensors that measure relative values. Other sensors, such as a limit switch (which may be used to signal that a robot appendage has reached a predefined position) or a pressure switch, indicate when an event occurs. This type of sensor can be used as a counter, as with a gear tooth sensor in a transmission. In this case, the sensor would count the number of gear teeth that rotate past the sensor, and that information would indicate the distance the robot has traveled.

The most sophisticated sensor supplied in the kit is the CMUCam2 Vision Sensor, a camera and micro-computer processor for robot vision designed by engineers and scientists at Carnegie Mellon University. In one configuration, the camera can be programmed to capture an image and determine the location of a specific color in the view.

The camera software provides output signals to drive the robot toward a specific color, such as the target above the goal. Camera output can also be used to trigger robot functions, such as launching balls when the target is in a predefined location in the camera's field of view. The camera is supported with software (also included in the FIRST Kit of Parts) that was written by the camera designers and augmented with a collection of programs written by FIRST volunteers and sponsors.

↑↑ The kit includes a camera that is capable of finding the center mass of a specified color. Servo motors are attached to the camera frame. Software on the camera's circuit board drives the motors continually to center the chosen color in the middle of the camera's field of view.

# ⇄ Putting the Pieces Together

Although all teams begin with the same set of components, each robot is a unique solution to the problem. In addition to strict requirements for weight (120-pound [54.4-kg] maximum) and size (5 feet [1.5 m]), the robots must meet stringent safety standards. The short time to design, construct, and test the robots requires the team to be efficient with their time and resources.

All work on the robots must be completed during the six-week period that follows the unveiling of the year's game during the first week in January. The robots must be completed by mid-February, at which time they are crated and shipped to the regional competitions. All teams must meet the shipping deadline to enter a regional competition, with no exemptions made.

Mimicking many real-world projects, the competition is meant to challenge teams with a hard problem to solve, a time frame that is too short, and limited resources. The quick pace and enormity of the endeavor energizes participants who rely on their creativity and ingenuity to accomplish much in a short period of time. The experience is often referred to as "hard fun" – a combination of terms that reflects a common attitude of participants who work hard on a fun project.

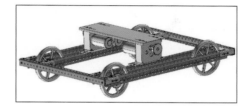

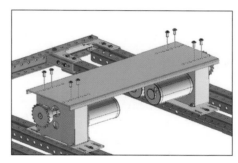

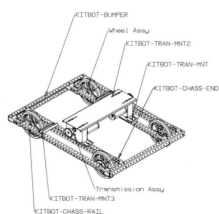

KITBOT-BUMPER
Wheel Assy
KITBOT-TRAN-MNT2
KITBOT-TRAN-MNT
KITBOT-CHASS-END
Transmission Assy
KITBOT-TRAN-MNT3
KITBOT-CHASS-RAIL

⇈ Teams learn by doing as they design and assemble their robot. Because the electronics, pneumatics, motors, and battery are specified by FIRST rules, all teams use the same power system. It is their challenge to find clever ways to use that power best.

# SECTION 01

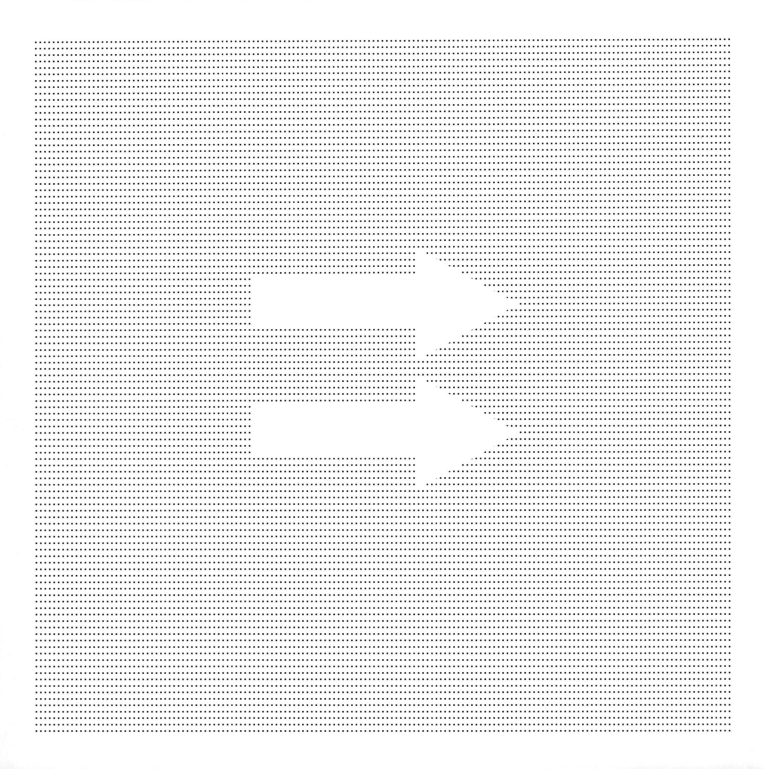

The Delphi Driving Tomorrow's Technology Award celebrates an advantageous machine feature that demonstrates any aspect of engineering elegance, including design, wiring methods, material selection, programming techniques, and unique machine attributes. The machine feature must be integral to the robot's function and must make the team more robust in competition.

Delphi is a global supplier of mobile electronics and transportation systems, including power-train, safety, steering, thermal, controls and security systems, electrical/electronic architecture, and in-car entertainment technologies.

As a founding sponsor of FIRST, Delphi has been instrumental in its growth and success. Generously supporting regional events and many teams, Delphi has given young people the opportunity to learn from and play with the 'pros' in the engineering world. Its long-term support of FIRST has also given its engineers the exciting, re-energizing opportunity to help students learn engineering skills, as well as form lasting connections with them from high school through college, internships, and into the workforce. The former Chairman and CEO of Delphi has served as a FIRST board member.

# Delphi Driving Tomorrow's Technology Award

# BUILDING A CHAMPION, ONE IMPROVEMENT AT A TIME

## RAIDER ROBOTIX'S UNDEFEATED SEASON OF REGIONAL COMPETITION

FIRST games are designed to be challenging, yet they still include an element of luck to keep them interesting. With a blend of skill and fortune determining the matches, every team has a chance of winning. Some teams, though, rely more heavily on skill than on luck to be successful.

Team 25, Raider Robotix, from North Brunswick, New Jersey, competed in two FIRST regional competitions, guided by the team motto, "Nothing is beyond our reach." The words rang true: the team was undefeated at both competitions and racked up a match record of 31-0. Along with the "champion" mantle, they were recognized with engineering awards for their design and flawless style of play. The judges at the Las Vegas Regional Competition said the team's robot had "numerous design improvements that set it apart from the others in the field…the robot proved to be able to play the game exceedingly well, scoring at will."

Team 25's 2006 machine was the latest in a long chain of robots, and benefited from years of improvement. The team has standardized some aspects of their robot, polishing their designs with incremental improvements to each legacy system. Three components they fine-tuned over the years—a modular gear-driven drive train, servo-powered brakes, and a camera-guided shooting mechanism—helped the team dominate the 2006 season.

## ⇄ Drive System: Improving Performance Since 2000

After reviewing the game, Team 25 knew that a powerful drive train would be key to scoring and to defending against other robots. They knew exactly what drive train would be ideal for the competition: the basic one used by the team since 2000. In each of the six years since its original design, Team 25 has made gradual improvements to the drive train to tune its performance.

The 2000 drive train was designed to push its way around a crowded field—a task similar to the one in the 2006 competition. The 2000 model used two hand-drill motors to power a four-wheel drive propulsion system culminating with 6-inch (15.2-cm) carriage wheels. This system worked well, but did not exhibit the desired power. The design was modified for the 2001 season with the addition of two more drive wheels. This modification was an improvement, but still did not provide enough traction.

In 2002, the team replaced the carriage wheels with 6-inch (15.2-cm)-diameter, 3-inch (7.6-cm)-wide molded foam rubber wheels, and larger motors were used in place of the drill motors. To increase the contact with the floor, the wheels were ground down using a cylindrical grinding attachment on a lathe. Symmetrical grooves cut into the rubber added more traction between the wheel and the carpeted playing surface. With these changes, drive performance improved greatly, but still more power was desired.

For the 2003 season, the team again changed the drive-motor configuration to use two drill motors and two other motors on each side of the robot. They also decided to replace the chain drive with gears, which required the team to study gear-driven transmissions to understand gear ratios and gear meshing. The improvements made during this season finally provided the desired power and traction.

In subsequent years, the only additional improvements were to simplify the design by reducing the number of parts, and use wheels that had gears directly attached.

The 2006 drive train used four of the most powerful motors in the Kit of Parts. The drive train gears were incorporated into lightweight modules that could easily be removed if needed. This feature permitted easier maintenance.

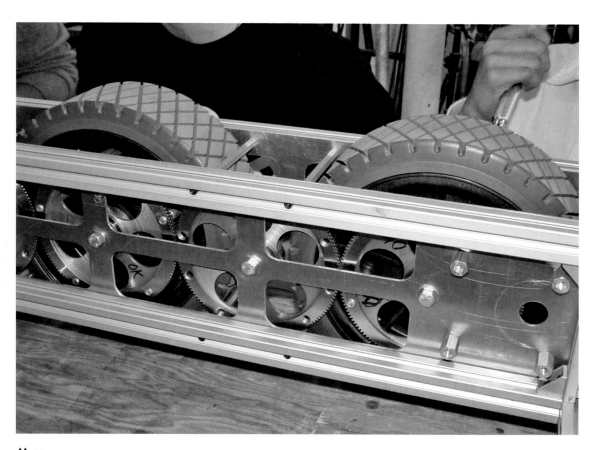

⇈ After several years of modification, a gear-driven system emerged. Here you can see the wheels and accompanying gears that drive them.

← An end view of one side of the robot, showing three drive wheels directly connected to the meshing gears, which are sandwiched between two metal plates.

↓↓ This drawing depicts the wheels and gears, emphasizing the gear teeth and the resulting gear ratios to move the robot at a top speed of over 9 feet (2.7 m) per second.

In addition to the powerful drive train, servo-motor-powered parking brakes were used in the propulsion system to prevent the robot from being moved by opponents. The primary transmission gears of the drive train had small holes into which spring-loaded steel pins could slide. The pins were designed to allow the brakes to be set even if the holes in the gears weren't perfectly lined up. The brakes also made it more difficult for an opponent to push Team 25's robot out of scoring position.

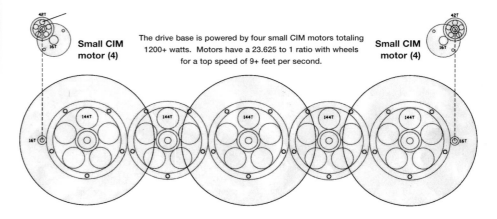

Small CIM motor (4)

The drive base is powered by four small CIM motors totaling 1200+ watts. Motors have a 23.625 to 1 ratio with wheels for a top speed of 9+ feet per second.

Small CIM motor (4)

↑↑ Four of the modular transmission components are seen here, stacked up and waiting to be installed in the robot.

→ The drive-train gears were incorporated into modules that are inserted through the wall of the chassis.

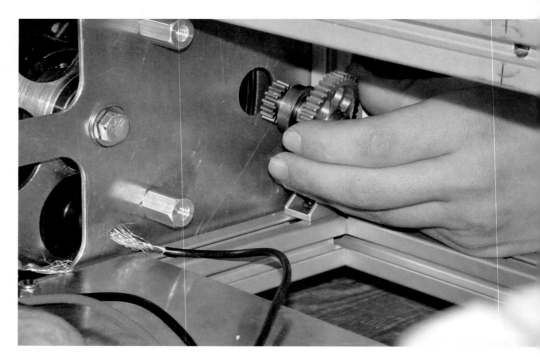

## ⇄ Designing the Perfect Wheel

⬇⬇ The servo-driven, spring-loaded brakes are mounted just above each CIM motor, as shown at two different angles. When these brakes are applied, the drive is disabled.

To compete offensively, Team 25 designed a camera-guided shooting mechanism mounted on a turret that was capable of rotating 300 degrees. Preliminary planning determined that a low center of gravity was vital for stability, and thus the shooting mechanism could not be mounted high on the robot. However, a low shooting device left open the possibility of shots being blocked by opposing teams. To remedy this situation, the team designed a shooter at mid-height. The shooting mechanism, tilt angle mechanism, and ball delivery system were all located on a single axis of rotation.

Like the drive system, the shooting system evolved over time. But unlike the six-year drive-system development, the shooting system evolved in just six weeks.

Computer-aided design (CAD) software helped the team visualize different configurations for ball launching. The original design used two vertically mounted wheels that launched a ball that was fed between them. When tested as a prototype, this design did not perform as well as expected.

The design evolved to be a single shooting wheel in proximity to a plate that would provide the pinch point to propel the ball. The plate also created backspin. Backspin was desirable because it increased the ball's travel distance and improved accuracy. The single-wheel shooter performed much better than did the dual-wheel shooter. With only one wheel and one drive motor, the weight of the mechanism was also greatly reduced.

The first generation of the shooting system was constructed using the same foam-rubber wheel used in the propulsion system. Testing revealed that when the wheel rotated at the required angular velocity to shoot a ball, the foam rubber separated from the wheel's plastic rim. The foam rubber wheel was replaced with a 10-inch (25.4-cm)-diameter pneumatic tire. The tire's edges were ground flat for better contact with the foam balls. While testing the pneumatic wheel, the team discovered it wasn't balanced, which led to significant vibrations when the wheel was rotated at launch velocity. To eliminate vibrations, a custom aluminum wheel was machined and fitted with high-speed, sealed bearings. This wheel was statically balanced with washers to ensure that it performed as smoothly as possible.

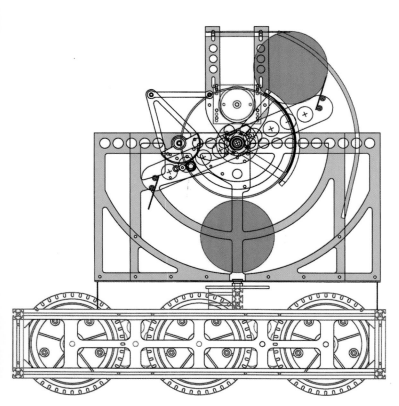

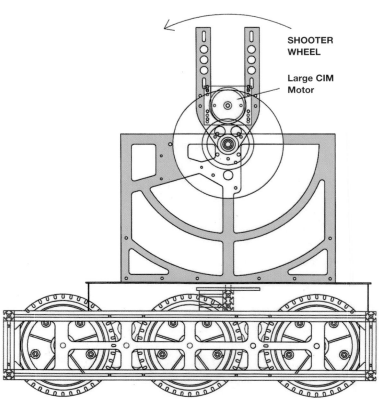

**SHOOTER WHEEL**

**Large CIM Motor**

⇈ A conceptual side view of the robot shows the combination of the tilt, shooter mechanism, and ball delivery functions on a single axis.

⇈ This conceptual side view shows the placement of the shooting wheel and the motor used to power it.

⇉ The finished shooting mechanism very closely resembles the conceptual view. The custom aluminum wheel can be seen clearly.

# ⇌ Shooting to Score

A motor-powered, tilt-control device was built to aim the shooting wheel into two preset positions for close and far shots at the goal. These angles could be preset for the autonomous mode, or set manually by an operator when in drive mode. This system, along with the camera-guided shooter, rotated on the gear-driven turret while tracking the goal.

This paddle-wheel design loaded two balls per second when operating at the maximum motor speed. This rate provided enough time for the shooting wheel to recover its angular velocity between shots.

To load balls into the shooting device, two arms fitted with scoops rotated around a motor-driven axle. These arms collected balls from the feed mechanism and passed them over the shooter wheel. The balls were dropped between the shooting wheel and the pinch plate.

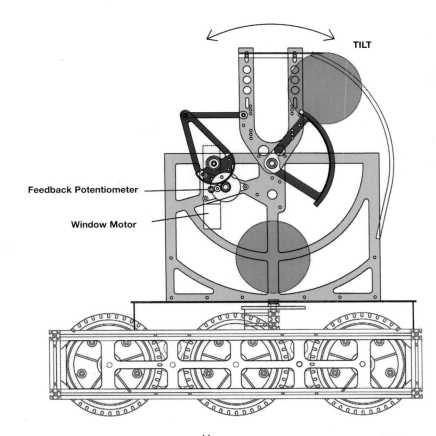

TILT

Feedback Potentiometer

Window Motor

⇈ This drawing depicts the functions of the tilt mechanism, including the motor and actuators that adjust it for close or far shots on the goal.

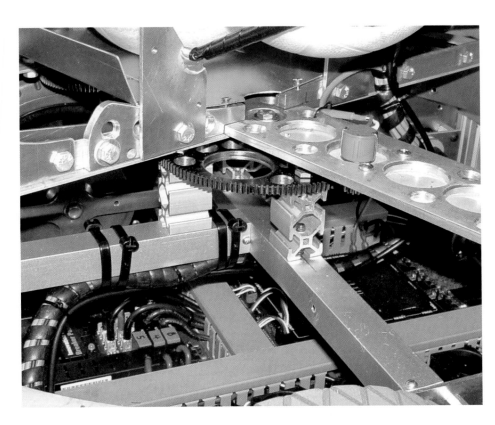

⇄ The tilt control and camera-guided shooting mechanism all rotate on a gear-driven turret.

➡ A line of balls is fed down to the scooped arms, which bring balls over the shooting wheel to be pinched by a back plate.

⬇ A motor powers two arms equipped with scoops to transfer the balls into the shooting mechanism. This drawing shows the placement of the motor and axle that control the arms.

## ⇉ "Nothing is Beyond Our Reach"

Team 25 continued to live up to its motto at the FIRST Championship. Its winning streak continued throughout the championship, with their only loss in qualifying rounds coming in its final match. The team went on to be a finalist in the competition, playing in front of 15,000 robot fans.

The team's robot continued to draw the attention of the championship judge panel as well, which awarded the team the Delphi Driving Tomorrow's Technology Award in recognition for engineering and building one of the best solutions. The overall design was noted for its simplicity, effectiveness, and reliability. The single-axis scoring system could effortlessly shoot balls into the three-point upper goal and did so match after match.

Team 25's success resulted from years of experience and continual improvement. With its openness to new ideas and ability to improve designs, Team 25 performed at levels far above others, embodying their motto "nothing is beyond our reach."

⇇ The completed robot found success in the evolution of its components.

# THE ROTARY BASKET SORTER: FROM IDEA TO REALITY

## A BEE HIVE FOR BALL STORAGE

Robot design usually begins with brainstorming to generate lists of possible solutions, in hopes that the best solution will rise to the top. A successful team encourages its members to apply their knowledge and not to be afraid to present ideas. When many designs are brought to the table, creativity blossoms, and even more avenues are opened for development and enhancement.

Team 33, of Auburn Hills, Michigan, was able to prioritize needs and to explore various design options before finalizing plans to build a robot. The end result of its brainstorming effort was an imaginative way to store, organize, and regulate game balls. This team, known as the Killer Bees, constructed a machine that gathered balls in a swarm-like fashion, much like a swarm of bees returning to a nest.

↑↑ A side view of Team 33's robot shows the
final design that included ball collection, storage,
and sorting capabilities.

## ⇄ Brainstorming: What Do We Need?

Team 33 determined while brainstorming that it wanted a robot that could hold more than 20 balls. Having that many balls on board meant the team would have the capability to score numerous times. Also, by controlling so many balls, they would be able to keep balls from their opponents.

Other elements of its plan included a design for loading the balls from the human player, a means to collect balls from the playing field floor, an ability to score in the high goal, and a way to mount the ramp at the end of a match. After reviewing this list of functions, the team determined that a ball storage and delivery system was a major component in most game strategies. The robot would have to be able to hold a large quantity of balls, handle a continuous feed of these balls, adapt to the different properties of each ball, and, most importantly, avoid jamming.

With these criteria, the team began to add details to its design. Several highly desired abilities were identified, such as a storage capacity for 20-40 balls and the ability to deliver one ball at a time at a rate of two balls per second. The team considered a ramp design to organize the balls, but found that it easily jammed, was heavy, and was inconsistent in its ball-feed rate. Experimentation with large bins showed that this storage form often resulted in ball jams at the exits, so that idea was also abandoned. Single, vertical columns could provide a controlled feed delivery rate but were limited by the number of balls that could be stored. After considering different designs, Team 33 focused its energy on a large bin equipped with a sorting device.

## ⇄ It All Starts With A Prototype

A preliminary, small-scale prototype was constructed using cardboard and golf balls to test the viability of the storage and sorting system. The success of this prototype encouraged the team and led to the construction of a full-sized mock-up made of wood. The mockup allowed the designers to better visualize the geometric restraints they would encounter.

Computer-aided design software provided further analysis of the design; a model of the storage system determined that six balls could fit into a 12-sided basket structure. The entire structure was contained within the team's size restriction for this part of the robot. Since this storage method would satisfy the desired characteristics, work began on fine-tuning the details of the mechanisms to support the storage cylinder.

⇈ A model was constructed from cardboard to arrange golf balls, which determined how to best arrange the foam balls in a storage container.

⬆⬆ A full-size wooden prototype of the storage bin was constructed to test the movement of a preliminary sorting plate and its operation with the foam balls.

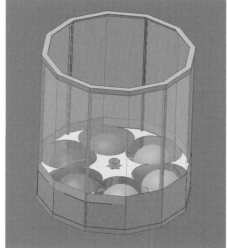

⬅ An AutoCAD drawing was made to size the bin and the 12 sides that would radially store the foam balls.

⇈ Looking down into the final bin provides a view of the wedge-shaped rotary basket sorter.

## ⇄ Advancing to the Final Design

The final design soon came to be called the "rotary basket sorter." While not exactly a poetic name, the title did define the system functions. The structure was constructed using a lightweight aluminum channel frame. The lower section of the basket was divided into six chambers by clear, cellular Lexan plates, while the upper section remained open. The transparent material for the dividing plates and the outer skin of the basket was chosen so that drivers could see the balls in the sorter system.

Balls could be loaded into the basket by the human player or automatically collected from the playing field floor. A sorter blade was fabricated from aluminum to sort the balls into each chamber. Team 33 determined that a blade height of two inches (5.08 cm) would achieve optimal stirring and create the least amount of drag. The skin of the storage basket revolved with the inner chamber dividers to reduce friction and decrease the number of moving parts that had to be manufactured.

Understanding that it all adds up, the team set a design limit for the sorter assembly's weight at 10 pounds (4.5 kg). This low weight meant, the sorter could be easily and quickly rotated. A single motor rotated the basket, minimizing weight and power draw. As the basket rotated, balls were directed at a controlled rate to the shooting mechanism. The weight restriction during the design phase allowed for weight allocations to other robot components such as the drive train and shooting mechanism.

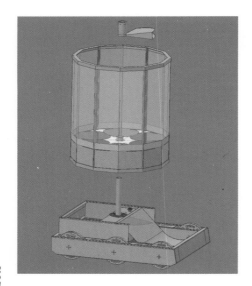

⇈ This AutoCAD drawing shows how the chassis, bin, and rotary sorter fit together. The Lexan sides of the bin allow the drivers to see the balls inside.

## ⇄ Coordinating Actions

A series of tests helped coordinate the robot systems. During the design phase, a digital tachometer and data acquisition recorder were used to measure the shooting wheel's velocity restoration time following a shot. This time interval was important because it established the optimum rate for feeding balls to the shooting mechanism.

The optimum feed rate of the balls from the chamber to the shooting device determined the gear ratio for the motor that drove the basket. The basket's rotational speed was designed to deliver just more than two balls per second when the drive motor operated at peak power. This speed would allow the robot to shoot approximately 10 balls in less than five seconds—a rate that met the team's launch-rate criterion.

The final sorter stored up to 40 balls, which also met the team's design goal. The sorting mechanism rapidly sorted and serialized the balls and controlled the feed rate. By varying the distance between each ball fed to the shooting system, the balls were only delivered to the shooter when it was capable of firing. Coordinated motion ensured that the systems operated as designed.

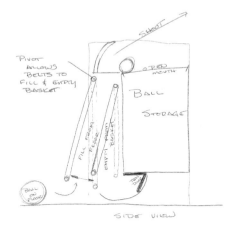

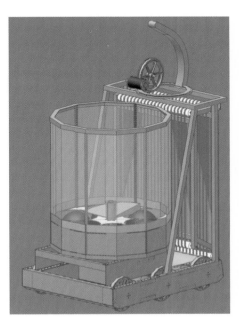

↑↑ A side-view sketch of the robot describes the pivoted ball-gathering device that collects foam balls from the playing field floor.

↑↑ A shooting wheel is seen at the top of the complete robot in this AutoCAD drawing. Balls that came through the sorter mechanism are sent up to this firing device via a vertical conveyor belt system.

↓↓ When the storage bin is filled to capacity, more than 40 balls can be stored from either the human player or the ball-collection device. These balls can then be shot, or saved in the bin to prevent opponents from acquiring them.

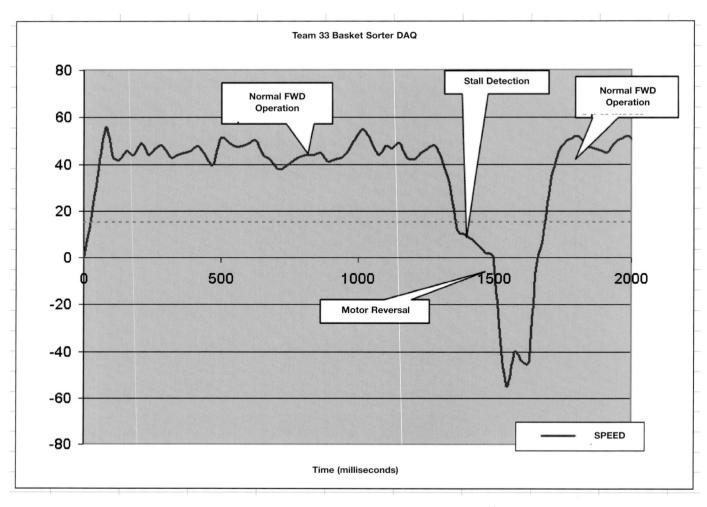

Team 33 Basket Sorter DAQ

Normal FWD
Operation

Stall Detection

Normal FWD
Operation

Motor Reversal

SPEED

Time (milliseconds)

⬆️ A graph of sorter speed versus time shows
normal motor operation, the occurrence of a
jam, motor reversal to correct, and the return to
normal operation.

## ⇄ Fixing It In Software

The rotary basket sorter was designed
to prevent ball jams, but because the
foam balls were compliant a jam could
happen at the most inopportune time.
After doing all they could with the physical and mechanical system to prevent
jams, the team turned to software to
correct the sorter when a ball jammed
the device.

A 360-degree rotation sensor was
connected to the sorter's axis. If the
sorter's rotation rapidly decelerated, as it
would when a ball jammed the mechanism, the sensor would detect the
change in condition. The robot controller
software was adapted so that it would
recognize these jams, reverse the direction of rotation to free the jammed ball,
and then resume as before.

A Java-based application was used
to collect and graph this data directly
from the robot. This graphical represen-
tation of the feedback signal showed
when a ball became jammed, the
reaction of the software to reverse the
motor direction, and the continuation of
the normal sequences when the ball
was cleared.

With this software correction in
place, operators did not have to manually correct ball jams. When the system
had been fine-tuned, the entire hopper
of 40 balls could be reliably unloaded
without risk of damaging balls or robot
hardware.

## ⇄ Unique Solutions: from Prototype to Final Design

From the smallest scale prototype to the final robot, Team 33 focused on the original goals for its robot: storing many balls and delivering them linearly at a controlled rate. The final design was unique, effective, and efficient on the field.

In addition to their creative mechanisms, Team 33's designers also applied analysis and programming logic to advance their design. The careful calculation of system timing and basket geometry produced a machine that could collect and store scores of balls and fire them at the goal in rapid succession.

The team demonstrated creativity not only in the mechanical design of the robot, but also in modifying the control software to correct for inconsistencies that it couldn't physically control. Such unique thinking demonstrates the value of diverse ideas at each phase of the design process.

⇄ Just as the team originally designed, the final robot could store many balls and had the ability to deliver them in a linear fashion at a controlled rate. The playing field proved to be no match for Team 33.

# RELIABLE, POWERFUL, ADAPTABLE: ATTRIBUTES OF STRONG DESIGN

## KEEP GOOD IDEAS AT THE READY

Although the actual construction period for the FIRST Robotics Competition is six weeks, teams are free to test ideas throughout the year. Sometimes teams will use the time between competitions to re-examine ideas that surfaced during the previous competition season but were not implemented. Progressing from an idea to a working design sometimes just takes a bit longer than six weeks.

During the 2004 FIRST Robotics Competition season, Team 100, from Woodside, California came up with a great idea for a versatile propulsion system, but there was not enough time to transform that idea into a working system. Rather than scrap the idea, the team stored the design on the chance they could find a use for it in a future competition. For the 2006 FIRST Robotics Competition, the team resurrected the stored design and converted it into a great competitive advantage.

## ⇄ Selecting a Drive Train

Team 100 began with a desire to build a robot that could easily climb the ramp and have strong defensive characteristics. Other robot functions would complement the drive train, so this system was the first component designed. In addition to the performance characteristics, the entire mechanism had to be simple enough that it could be built using the limited tools and resources available to the team.

Three different ideas were presented for the drive system. A swerve drive was proposed to allow omni-directional movement, but this system was too complex to build, given the team's capabilities. Also, a swerve drive system was not ideal for the type of robot needed for the 2006 FIRST game.

A two-speed, six-motor, six-wheel drive propulsion system was also suggested, as it would be fast and have the power to force its way around the field. This design was also considered to be too sophisticated given the available resources.

The third alternative was to build the drive system from the material provided in the FIRST Kit of Parts. This would be easy to put together, but it also meant that many other teams could be using the same drive and there would be no advantage gained.

The final design merged two of these ideas. A two-speed transmission would be combined with parts in the kit to drive four wheels. The transmission gearboxes would be constructed and mounted on the kit frame. This would create a powerful, multispeed drive system mounted on the easy-to-assemble frame.

The drive system called for six of the most powerful motors supplied in the Kit of Parts to provide a propulsion edge over teams using a four-motor propulsion system.

Transmissions would shift the system into a low-speed, high-torque mode or a high-speed, low-torque mode as required. The high-speed mode would be ideal when playing offense, to allow the robot to outmaneuver other robots. In defensive mode, the high-torque option would provide an advantage with a stronger ability to block opponents. The high-torque option would also improve the robot's ability to climb the ramp. Using six motors allowed for a greater range of speeds without compromising the rotational torque applied to the wheels. The concentration of motors also created a low center of gravity for the robot.

⇄ Custom gear boxes, combined with the kit frame, provided Team 100 with an easy-to-build chassis powered by a shiftable drive.

# ⇄ Six-Motor Drive: An Evolutionary Design

The drive train on Team 100's 2006 robot resulted from two years of tweaking. The team first proposed using a two-motor, two-speed transmission in 2004, but didn't implement it. However, the proposal sparked interest in a dual-speed option that would be reliable and advantageous. Although a shifting mechanism presented complications to the drive system, it was believed to be worth the effort.

Following the 2004 FIRST Competition season, a preliminary version of the dual-speed drive system was built. Though large, bulky, and heavy, the system demonstrated reliability and utility. Throughout 2005, the team continued to fine-tune the dual-speed transmission concept.

A second revision to the transmission used gears cut from steel spur gear stock. Transmission components were upgraded to handle the high loads placed on the system. The shifting mechanism's gears were heat-treated to accommodate the forces from shifting and changing direction and to prevent failure.

The gearbox housing was fabricated from a 1/4-inch (0.635 cm) -thick aluminum plate, into which ball bearings were pressed to hold the steel drive shafts. The actual shifting mechanism used a form of "dog-shifting," where a metal dog with protruding teeth slid along a hexagonal shaft to mesh into pockets that were milled into a gear riding along the same shaft.

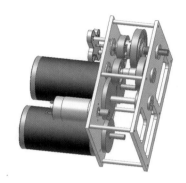

⇒ The second revision of the transmission was sketched and drawn in AutoCAD. Two CIM motors and one Fisher Price motor, coupled with gears that were housed in an aluminum gearbox, powered this version.

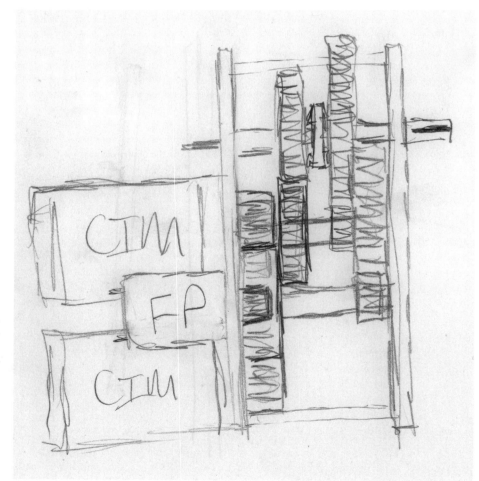

## ⇉ Third Time is the Charm

For the 2006 FIRST competition, a third revision was made to the transmission design. This time, the system was rearranged and simplified so the gears all had the same pitch. Also, the number of gears and other mechanical parts was reduced, lessening the chance of breakage and reducing the overall size of each mechanism.

Gear ratios were calculated to match the speeds of the motors, and speed, torque, and energy-loss estimates were also calculated. Using these data, along with analysis and tests to determine the friction coefficient between the wheels and carpeted floor, the team finalized the gear ratios. This third revision of the transmissions became the final design used on the 2006 robot.

When the final gearbox components were fabricated, each unit was individually assembled and tested. Each transmission was operated at a slow speed to ensure correct meshing of the gears and that no jams occurred. This test was followed by a high-speed test to break in the gears and check for stability.

Each CIM drive motor was tested for faults that could occur when engaging the transmission between gear positions, with the Fisher Price motor added to the transmission after all individual tests were completed. The robot was then run as a whole and performed almost exactly as predicted in the calculations.

↓↓ A complete transmission, with CIM and Fisher Price motors placed opposite each other, lies across elements of the chassis frame.

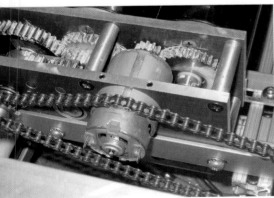

↑↑ A close-up view of the Fisher Price motor shows how it is connected with gears to the CIM motors.

↓↓ ⇉ The third and final revision of the transmission rearranged the position of the Fisher Price motor and simplified the internal gears.

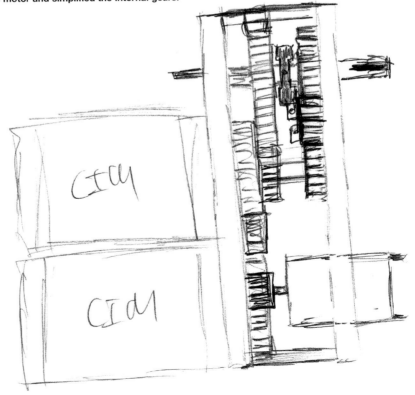

## ⇉ Adding Wheels To Be Ready To Run

A six-wheel skid steer layout was selected to transfer the power from the motors to the playing field floor. Six-inch (15.2-cm)-diameter caster wheels wrapped in wedge-top conveyor belt tread were fashioned to provide extra grip.

Between the transmission and wheels, a chain and sprockets provided the needed reduction to optimize performance. To ensure reliability, sprockets had to be at least 50 percent covered by the driving chain. Delrin rollers were added to guide the chain around the sprockets and to keep the chain within the frame of the robot, protecting it from outside interference.

⇊ A team member works on the drive wheels, which are wrapped in a special tread for additional traction on the playing field floor.

⇄ This side view of the robot reveals the white Delrin rollers which guide and protect the drive chain.

With two years already invested in the transmission and drive design, construction flowed effortlessly. A complex system was developed, fabricated, and thoroughly tested well within the six-week build time. The powerful and reliable drive train helped the robot play a significant role in complementing many alliance partners.

⇊ There is little difference between the intricate AutoCAD drawing of the robot and the finished product, whose success is based on its reliable drive.

# TEAM 357

# DESIGNING A BETTER WHEEL

## ROLLING OVER THE COMPETITION

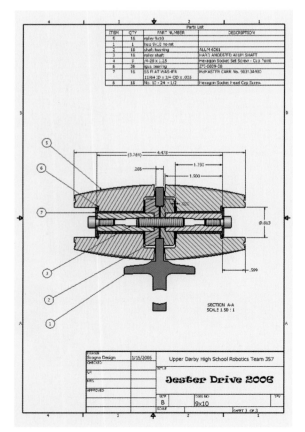

| Parts List | | | |
|---|---|---|---|
| ITEM | QTY | PART NUMBER | DESCRIPTION |
| 5 | 16 | roller 9x10 | |
| 1 | 1 | hub 9x10 no rat | |
| 2 | 18 | shaft bushing | ALUM 6061 |
| 3 | 18 | roller shaft | HARD ANODIZED ALUM SHAFT |
| 4 | 9 | #4-20 x .25 | Hexagon Socket Set Screw - Cup Point |
| 6 | 36 | igus bearing | ZFI-0609-06 |
| 7 | 16 | SS FLAT WASHER 11/64 ID x 3/4 OD x .035 | McMASTER CARR No. 9031 34 400 |
| 8 | 18 | No. 10 - 24 - 1/2 | Hexagon Socket Head Cap Screw |

(3.781)   4.478

.208   1.730

1.500

Ø.813

SECTION A-A
SCALE 1.50 : 1

.599

| DRAWN Scogna Design | 1/15/2005 | Upper Darby High School Robotics Team 357 |
|---|---|---|
| CHECKED | | |
| QA | | TITLE |
| MFG | | **Jester Drive 2006** |
| APPROVED | | |
| SIZE B | DWG NO 9x10 | REV |
| SCALE | | SHEET 3 OF 3 |

The Royal Assault Robotics Team 357 from Drexel Hill, Pennsylvania, solved the frictional dilemma for four-wheel drive robots by designing and manufacturing a drive system that used 18 rolling surfaces on each of four wheels. Each 10-inch (25.4-cm) wheel has nine rollers mounted at a 45-degree angle along the wheel edge. The rollers are split so that each side rotates independently. A shaft runs through each of the nine spokes of the wheel hub, with the rollers mounted on the common shaft. The resulting drive system, called Jester Drive, is one that allows motion in any direction, since the wheels translate and rotate simultaneously. Translation results when the rollers rotate on their own axes and rotation is achieved when the wheel spins on its axis.

This omni-directional motion is possible because wheel rotation produces a force that is perpendicular to the roller axis. For sideways motion, adjacent wheels on each side are driven in opposite directions, the resulting force being a side force on the robot. Since the robot orientation is independent of the direction of motion, under this condition the robot appears to float over the playing surface. Because opposing corners of the robot have rollers orthogonal to each other, the robot must be driven to purposefully move sideways.

This wheel design, known as the Mecanum Wheel, was patented in 1973 and has been commercially developed by Aitrax, Inc. as a means for increasing fork lift maneuverability. Team 357 partnered with Airtrax, Inc., to apply this technology to their robot drive system for the FIRST Robotics Competition.

Team 357 designed its new wheel using Autodesk Inventor CAD software and developed the necessary manufacturing techniques to produce the wheels' rollers and hubs.

⇇ This Autodesk Inventor drawing illustrates the components of each roller. The roller includes two sections of urethane, mounted on a common shaft and supported by a bearing pressed into the wheel hub.

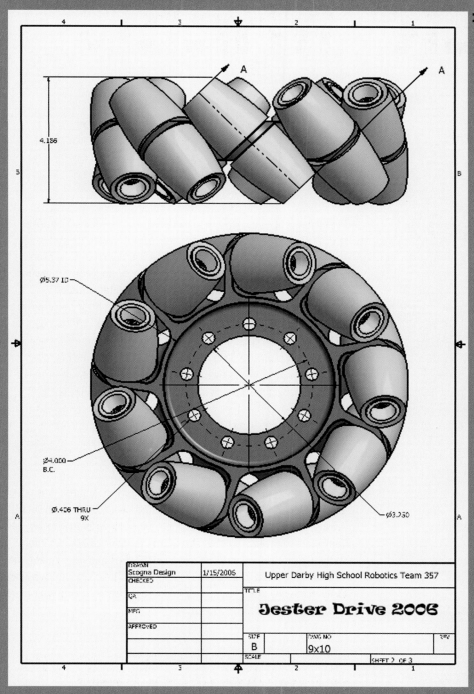

DRAWN Stagna Design 1/15/2006
CHECKED
QA
MFG
APPROVED
Upper Darby High School Robotics Team 357

TITLE

**Jester Drive 2006**

SIZE B
DWG NO 9x10
SCALE
REV
SHEET 2 OF 3

## ⇄ A New Wheel for Robot Propulsion

Jester Drive is a unique robot wheel design that provides two distinct advantages over conventional wheels: high traction for propulsion, and low friction when maneuvering. While a traditional wheel with a single rolling surface is most efficient in one dimension, a Jester Drive wheel has 18 rolling surfaces to create a wheel that efficiently rotates in two dimensions. With four Jester Drive wheels on a robot, omni-directional movement can be obtained.

⇄ Jester Drive is an advanced robot propulsion system that uses a cylinder of individual rollers as a driving wheel. Ten rollers mounted 45 degrees to the wheel's centerline ring the wheel, and rotate independently of the wheel.

# ⇉ Robot Maneuverability

Robot traction is a perplexing constraint for robot designers. While high traction is desirable in most situations, such as pushing another robot or climbing a ramp, that same traction can limit the robot's maneuverability. A common drive system on FIRST robots consists of two wheels on each side of the robot. All four wheels might be powered, or one might be powered on each side, with the other freewheeling. In either case, traction is important for the wheel to establish solid contact with the ground and to propel the robot in the direction the wheel is spinning. However, when the robot needs to turn, high levels of friction between all four wheels and the ground is not advantageous.

The maneuvering dynamics for a robot are much different from other wheeled vehicles. In a car, for example, the front wheels pivot and guide the vehicle through a turn. Most FIRST robots do not have pivoting wheels. Turns are accomplished by rotating each side of robot wheels at a different speed or direction, and this motion causes the wheels to skid sideways across the ground. In the extreme case when the sides are powered in opposite directions, a robot can potentially spin within its own footprint as the wheels slide across the ground. High levels of friction often impede the sideways slip of the wheels and restrict robot maneuverability. Overcoming the resulting sideways forces on the wheels during turns requires large amounts of power or creative design.

One solution to this problem is an omni-wheel, which consists of a series of free-turning rollers, mounted at 90 degrees to the diameter of a wheel. The closely spaced rollers form a nearly continuous, round surface along the outer edge of the omni-wheel. These rollers rotate if the wheel is subjected to side loads, as is the case when turning. Unfortunately, the free-wheeling rollers provide no resistance to side loads exerted by another robot or by an obstruction. In other words, another robot can easily push an omni-wheeled robot sideways.

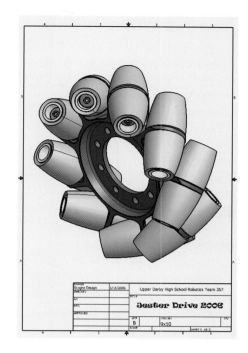

⇉ Three-dimensional computer models illustrate the offset angle of each set of rollers. The rollers are designed to form a continuous circular surface along the wheel's edge.

# ⇄ Molding Wheel Rollers

The first iteration of roller design consisted of making molds from cardboard tubes packed with modeling clay. The cardboard tubes were mounted on a lathe and turned to the correct mold shape. Urethane rubber was poured into the molds to form each roller. This process yielded inconsistent rollers which created a bouncy ride for the robot.

The most concentric roller was selected from this initial batch and used to make a separate mold out of silicone. The roller was placed in an aluminum cylinder and silicone was poured around the roller. This technique produced uniform rollers from a common mold.

The first rollers were made with urethane that became hard when cured, yielding rollers with little traction. A different formulation of urethane produced softer wheels, with higher degrees of traction. Though higher degrees of traction were obtained, the softer urethane was unable to retain its shape. These softer rollers deformed at their ends and resulted in a condition where the hard bearings inside the rollers were in direct contact with the ground.

The deforming roller tip problem was corrected with a dual-layer roller, where each layer had a different hardness. Urethane with a harder durometer for the inner core helped the roller retain its shape, and a softer urethane was used as an outer tread to produce better traction. The team used Autodesk Inventor to produce solid models of the new design. Stereo lithography was used to create three-dimensional models of the roller, and this model served as a template to make the silicone molds. A silicone mold of six inner cores and six outer tread layers was the final result of this process.

A dual-molding process began with an initial pour of the inner core using the harder durometer polyurethane. Plastic bearings, manufactured by Igus, Inc., were placed in the mold before the pour to provide a roller surface for each roller shaft. Before the mold was fully cured, the core was transferred into the larger diameter mold and the tread layer was poured around it. Because the inner core was not fully cured, the softer durometer outer layer bonded to the inner layer to produce a dual-layer roller.

⇄ Roller production begins with mixing the urethane for the molds. The final mixtures used in the design—a harder inner material surrounded by a softer outer shell—are the result of a series of tests to identify the optimum urethane hardness.

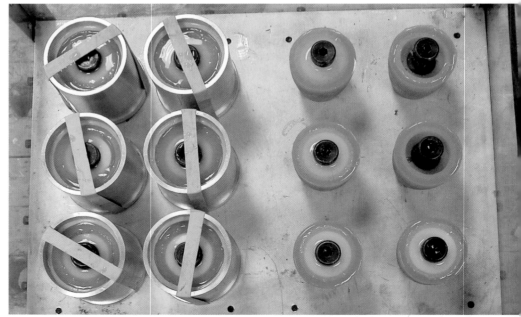

⇄ A dual-stage pouring process produces the rollers. In the first pour, a thin layer of pink foam leaves room between the inner urethane and the mold. The rollers are removed from the molds, the spacing foam is removed, and the second pour commences to create the composite rollers.

⇄ With two halves to each roller, and ten rollers on each of four wheels, 80 pieces are needed to outfit each of the robot's drive wheels.

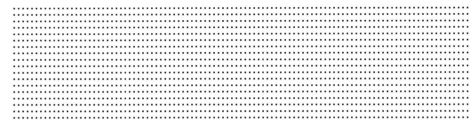

## ⇄ Hub Manufacturing

The hub design relied on computer-aided manufacturing techniques to produce the complicated shapes on which to mount the rollers. Having previously machined a 6-inch (15.2-cm), six-spoke wheel from a solid block of aluminum, requiring hours of flawless machining for each hub, the team realized that an alternate fabrication method was needed for the larger and more complicated wheel. Investment casting was ruled out because of the long lead-time and specialized tooling needed to make wax models of the hubs. Z-casting, a process of printing a three-dimensional model using a sand and adhesive mixture, was selected as the means to produce the molds.

The Autodesk Inventor model of the hub defined the shape, which was printed using Z-casting. The size of the wheel required that the hub mold be printed as four separate pieces assembled together to form the complete hub mold. Molten, aircraft-grade aluminum was poured into the composite mold to produce the hubs. The solid hubs were then machined, with the fins and mounting surfaces refaced and the axle holes resized, to ready the hubs for use.

⇄ The Jester Drive wheel hub is visualized using Autodesk's Inventor design software. Each spoke is offset 45 degrees to properly align the rollers.

⇈ The computer model of the hub is used with three-dimensional printing methods to create a mold for this shape. Aircraft-grade aluminum is poured into the mold to produce the hubs.

⇄ Light machining prepares the hubs for use in the Jester Drive. This machining is needed to correct mold imperfections and to achieve the low levels of tolerance needed to press bearings into the hubs.

# ⇄Improved Robot Performance

The sophisticated manufacturing techniques for the rollers and hubs, combined with skillful computer-aided design, produced a winning drive system for Team 357. The drive system had very high traction for climbing and pushing, was maneuverable in any dimension along the field, and could not be pushed by opposing robots. As detailed in an award proclamation for their work, the team "broke the mold in their tireless pursuit of reinventing the wheel, driving tomorrow's technology in any direction."

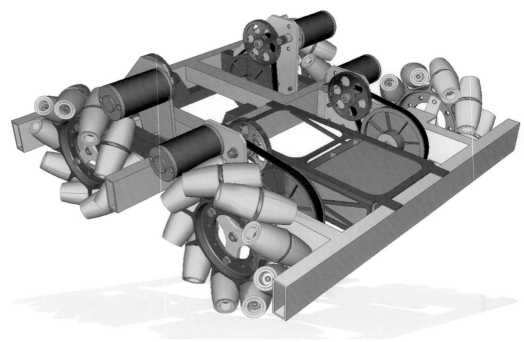

**↑↑ The wheels are one part of the drive train: a system that includes motors and transmissions to transfer power to the wheel. Because each wheel is controlled independently the robot can be commanded to move in any direction without changing its orientation.**

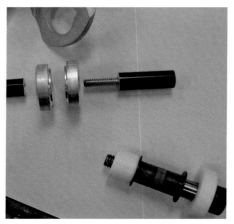

**↑↑ ⇄ The axle and bearings are key components that support the robot load and minimize friction. A close fit between components is necessary for the system to operate as designed.**

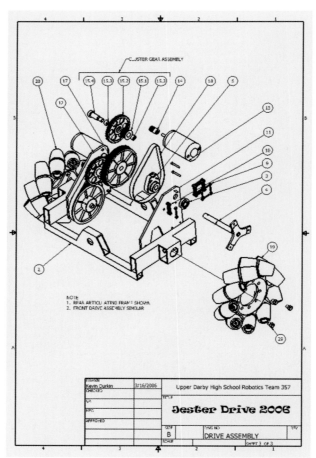

⇊ ⇒ Though partially masked by other robot components, the Jester Wheels are a distinctive attribute of the robot and illustrate a perfect balance between form and function.

# SOLVING AN ANTICIPATED PROBLEM

## CONVERTING OFF-SEASON DESIGN INTO REALITY

Team 467, of Shrewsbury, Massachusetts, got a jump on the competition by working as a team during the fall of 2005, before the 2006 FIRST game had even been announced. Classes were offered to sharpen the team's drafting and engineering skills. Also during this period, an "Introduction to FIRST" class was used to acclimate new team members to the world of FIRST.

During this time, the team decided that one of its priorities for the year would be to develop a reliable drive train. Ideas began to circulate about possible wheel configurations and power arrangements. Some of the configurations considered were a holonomic drive, a skid steer mechanism, and an independent four-wheel steering design. After much discussion, the team decided to pursue a tank-style propulsion system.

By December, discussion had moved to the transmission, and the team decided that gearboxes with worm gears would be the best option for a four-wheel drive robot. Because of problems in the past with uneven weight distribution over each wheel, and the negative impact of that condition on the robot's ability to steer straight, the team concluded that a suspension system would be needed on the robot.

## ⇄ Putting Practice to Work

When the game was announced in January, Team 467 knew that its design work would pay off and immediately began work on the proposed drive train. A prototype was fabricated from wood, supported with a combination of both rubber and omni-directional wheels. Each was powered by its own motor and gearbox, resulting in a four-wheel, four-motor drive system. Each gearbox, constructed with polycarbonate plate, incorporated double-threaded worm gears to propel the attached wheel.

Initial testing of the drive revealed that the polycarbonate boxes were too fragile and cracked as a result of the high torque produced by the worm gears.

To correct this problem, the gearbox structure was modified to consist of two C-shaped halves. A thicker polycarbonate plate, located closer to the drive wheel, contained the hex wheel shaft and worm gear. The other half was constructed of aluminum and nylon and supported the drive motor, motor shaft, and worm gear.

Aluminum covers were created to hold the bearings and motor shaft in place, and tensioners were added to keep the gears at a uniform separation and to improve how the gears meshed with each other. Following more testing, the double-threaded worm gears were replaced with a single-threaded worm gear of the same size, doubling the gear ratio and reducing the system's complexity.

⇊ **A working wooden prototype drives on a combination of rubber wheels, and omni wheels.**

⇊ **A close-up of one of the omni-wheels on the wooden prototype shows its connection to an individual gearbox and motor.**

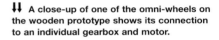

⬆⬆ A view-inside an aluminum gearbox shows the single threaded worm, attached to a drive motor and worm gear.

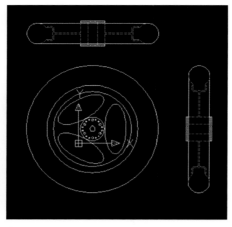

⇆ ⬇⬇ These drawings show the design of the custom-made wheel hub and wheel, which held both rubber and omni wheels on the robot.

Two different materials were used to construct the drive wheels. On the front end of the robot, omni-directional omni-wheels were employed to provide better steering control. The initial friction on the rubber rear wheels was very high and limited maneuverability. To reduce this friction and make the robot easy to maneuver, the rear wheels were covered with pool hose material. Custom hubs were fabricated to connect the wheels to the drive shafts.

The wooden plates in the prototype were eventually replaced with aluminum to improve performance and conserve weight. To further reduce the weight of the robot, a strong, lightweight, honey-combed composite material—constructed from Fiberglass and polycarbonate—was used for the back and center plates.

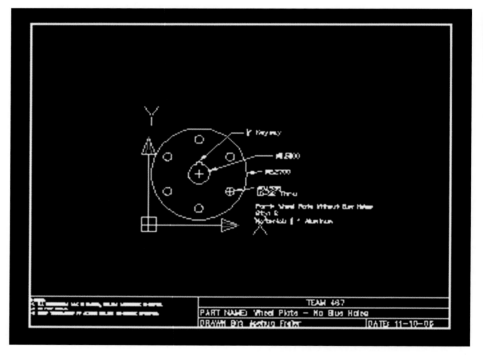

➡ Looking at the edge of one of the outer panels reveals the honeycomb pattern of the Fiberglass-polycarbonate composite. This material is very lightweight, and because of the honeycomb pattern, also very sturdy.

# ⇉ Maximizing Performance by Improving on Past Designs

To correct for past problems with uneven weight distribution over the wheels, Team 467 incorporated a suspension system into the robot. Possible suspension systems included a spring suspension and a powered pneumatic suspension. The spring system was abandoned because it was too complex to prototype and the pneumatic system was not pursued because of its excessive weight. Flexible chassis material was also ruled out because of a lack of structural support and limited possible movement.

The chosen suspension was a modular design. This system was originally tested with LEGOs. CAD drawings were created for each robot component. These models were refined and improved as the season progressed. The CAD drawings greatly aided the team's ability to work together and simultaneously design all aspects of the drive train and suspension system.

A bracket was placed on each side of the chassis centerline to support a suspension shaft. These shafts were mounted to the side panels, which consisted of the drive motors, gearboxes, and drive wheels. The shafts rotated in the brackets, and as the robot's wheels moved up or down the shafts slid through the mounting holes.

⇈ One of the brackets can be seen mounted to the edge of the center board. This bracket, along with others on the robot, encased suspension shafts while allowing them to slide and rotate.

⇈ A suspension shaft can be seen protruding from one of the side panels of the robot. Each side panel holds the wheels, motors, and gearboxes for one side of the robot.

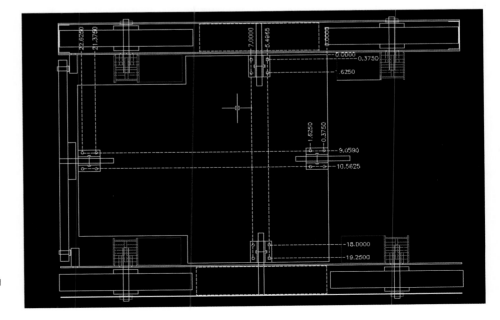

⇉ An overhead drawing of the chassis shows the positions of the brackets and shafts, located in the center of each of the four sides of the robot.

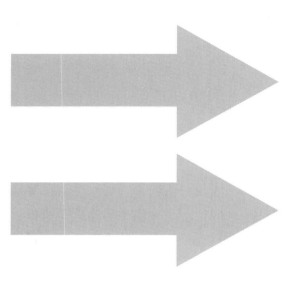

Each corner of the robot had a pivoting ball joint that allowed the frame to flex while connecting the robot's four sides. The joints had a quick disconnect mechanism, so they could be easily separated to remove a side of the robot for fast repairs. The suspension shafts connected the side panels to the center of the chassis, allowing the frame to be flexible and to have a range of motion. Originally, the suspension shafts were fabricated from steel, but the material was changed to titanium to reduce weight and the chances of bending and breaking.

⇉ ⇈ Pivoting ball joints are used to connect the four sides of the robot, allowing them to twist and flex as they move independently. Both a close-up view of the joint and a view of the connected robot can be seen here.

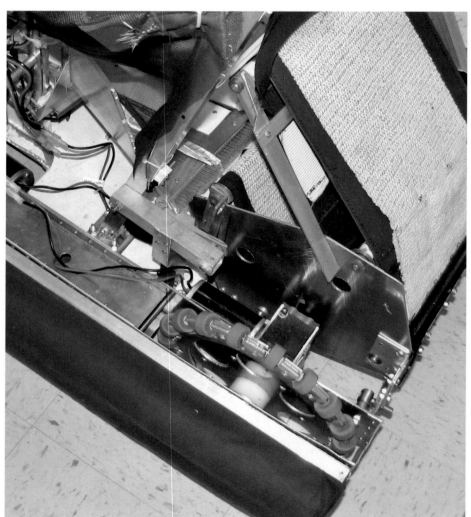

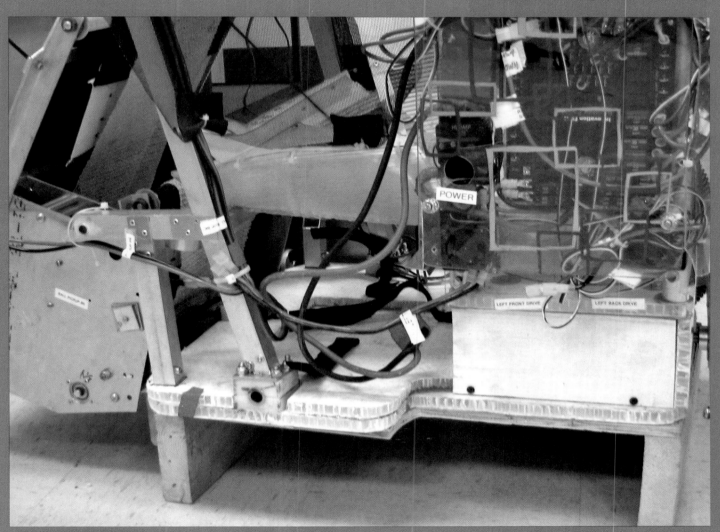

⬇⬇ **With the side panel disconnected, a component of the pivoted ball joint can be seen attached to the right panel.**

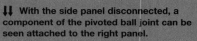

POWER

BALL PICKUP ARM

LEFT FRONT DRIVE

LEFT BACK DRIVE

## ⇄ The Advantage of Robust Design

Given the open competition field, Team 467 anticipated high-impact game play and designed a robust robot that was easy to repair. To allow access to internal robot components, the side panels on the robot were fastened to the chassis with quick-disconnect fittings at each corner. With the quick-disconnect fasteners, no tools were needed to remove the panels, thereby simplifying the repair process for internal systems. The side panels were made of aluminum prisms and connected to the composite material in the rear panel to provide a tough outer shell for the robot.

Although designed for a propulsion advantage, the suspension system was also a strategic benefit as it increased the robot's stability, especially when climbing the ramp and colliding with other obstacles on the field. The suspension also reduced the amount of shock transmitted to the robot's systems, thereby increasing overall reliability of the robot.

The front of the robot incorporated a ball-collection system directly in the chassis to prevent interference with the suspension system. With the use of quick-disconnect fasteners, the collector could also be easily removed from the robot for repair and maintenance. This type of modular design, in which entire systems can be easily removed from the robot, made it possible for multiple systems to be simultaneously repaired and aided the team's ability to be ready for each match.

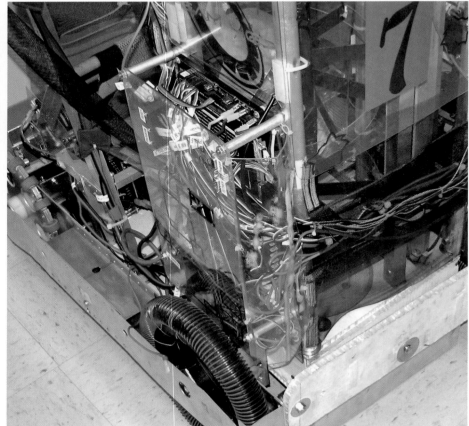

⇊ Aluminum side panels and a Fiberglass-polycarbonate, honeycombed composite rear panel comprise the outer structural elements of the robot. These can easily be removed for internal repairs.

⇈ Here is another view of the quick disconnect mechanism at a corner of the robot.

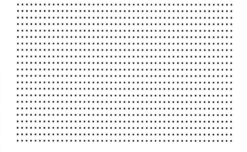

## ⇄ Preparation Enhances Performance

Team 467 benefited by anticipating its needs and preparing for the competition season before it arrived. Its foresight resulted in preliminary designs of a drive system and a suspension system before the competition game was announced. The FIRST Robotics Competition regulates that actual construction of the robot must occur in January and February of the competition year, but the rules do not restrict thinking to only this time period.

Team 467 took a gamble by designing a system before the robot's requirements were specified, but that gamble paid off. The 2006 FIRST Robotics Competition called for fast, high-torque robots, and that is the system Team 467 had ready. By starting the season with a drive system and suspension system ready to be built, the team could turn its attention to other design challenges. This design preparation, combined with year-round team development activities, allowed Team 467 to have a preliminary design when most teams were brainstorming.

⇄ The modular structural components and dual-wheel materials in the completed robot.

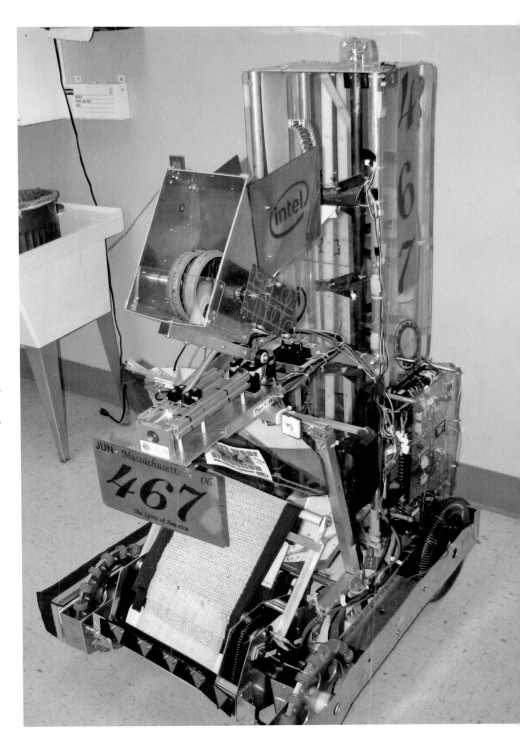

# MONOCOQUE CONSTRUCTION: USING A SHELL FOR STRUCTURAL STRENGTH

# ADAPTING A CLASSIC METHOD FOR ROBOTIC DESIGNS

FIRST Team 1501, from Huntington, Indiana, entered the competition with a limit on physical resources, but with an unlimited capacity for innovation. As a second-year team, the group faced a tight budget and only had a small machine shop. The team wanted to develop their own chassis, but with few supplies and no experienced welder, its options were limited. That limit, though, prompted creative thinking and led the team towards a creative construction method to build their robot frame.

One mentor, a former aviator, suggested the team look into monocoque construction. Monocoque is a construction technique that uses a structure's external skin to support applied loads. This method was used in the 1930's in aviation construction to strong structures while minimizing the weight of the aircraft.

Energized by the possibility, the team researched monocoque construction techniques to explore their viability in robotics. The team was surprised to learn that this "old fashioned" construction method is in fact the dominant automobile construction method used today. For most autos, pressed panels are welded together to form a rigid shell that bears all torsional and bending stresses. Windshields are often incorporated as an integral part of the shell to provide structural strength.

⬇⬇ **The final robot closely resembles the structural components of an airplane, with clean geometric lines and interesting material selection.**

## ⇉ Constructing a Robot Base

Support can be provided to a body internally or externally. The concept of frame construction is well understood because of its popularity. With frame construction, whether in a robot, in an automobile, or in another structure, an internal frame provides a large proportion of the structure's strength. External body panels commonly are hung on the frame, but the panels offer little role in structural strength.

The word *monocoque* is from the French words *single shell*, and this legacy provides insight into the construction technique. In monocoque construction, the external panels provide all of the structure's strength. The panels are rigidly joined together to form a shell. Any load on the structure is distributed and supported by the shell. The shape of the shell is designed to evenly distribute the load, avoiding any sharp corners or non-uniform thickness that would concentrate stress in one area.

The panels are usually constructed from thin metal sheets. The metal's resistance to tearing enables the panels to support shear and tension loads. When panels are bent, such as along their edges, they become strong enough to resist bending loads as well. The only load that is not supported is a compression load across the panel. To resist compression, structural stiffening can be added to prevent buckling. Additional strength can be added to any panel by folding a panel edge over itself to provide more structure along the structure's border. Team 1501 was motivated to use monocoque construction for two reasons. The structure had a high strength-to-weight ratio, and as such would help keep the robot within the weight limit. Also, the structure could be constructed with hand tools. Using a band saw, tin snips, a drill, and a rivet gun, the team could construct its robot base. The team confirmed its inclination to use monocoque construction after analyzing the weight, center of gravity, clearance, and stability of the proposed design.

The base was constructed as a series of components that were fastened together. For additional rigidity, the larger pieces were augmented with an internal arrangement of ribs and partitions to transfer any imposed loads across the entire aluminum skin. Reinforcing plates were added as locations on which to mount other components such as a motor and wheels.

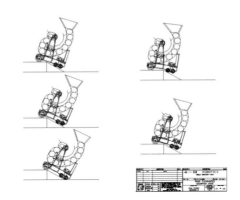

↑↑ An AutoCAD drawing shows the robot in different positions with respect to the ramp.

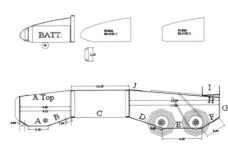

↑↑ ↓↓ AutoCAD drawings of the base and wing components are the first step of the monocoque design process.

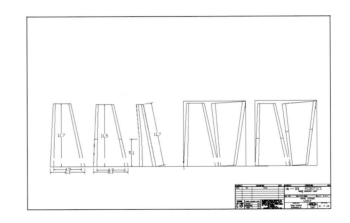

Team 1501 followed a simple process to convert sheets of aluminum into their base structures. First, the structural components were designed and drawn in two dimensions using AutoCAD software. These models were then printed in full size on paper. After cutting out the outlines on the paper, patterns were glued to 3/4-inch (1.9-cm) plywood.

The patterns were cut using a band saw to create form blocks. After the edges of the wooden form blocks were sanded, the blocks were placed on 0.04-inch (10-mm) aluminum sheeting and the edges were traced using a felt-tip pen. The team was careful not to use a scribe to trace the shapes, as the marks could become stress risers and cause cracks or failure points when in use. The pattern was traced again, this time 3/4-inch (1.9-cm) out from the form edges to allow for flanges that would be bent around the form.

After the wooden forms were removed, the aluminum patterns were cut along the outer lines, using either tin snips or a band saw. Relief notches were cut at corners and other places where the flanges would meet, and the entire aluminum sheet was deburred to remove stress risers. Perpendicular cuts were then made from the outer edge to the inner outline to create tabs that could be bent around the form edges.

⬇⬇ Life-size paper plots were made of components to be fabricated, then glued to plywood and cut out.

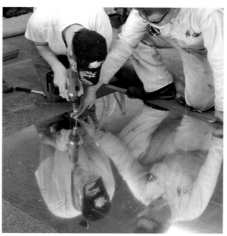

⬆⬆ The shape of the form block has been drawn on the aluminum, and 3/4-inch (1.9-cm) flanges can be seen traced around these.

⬆⬆ A wooden form block is seen atop a sheet of aluminum as it is traced using a felt tipped pen, and then retraced, to create a flange pattern.

A second set of form blocks were created by tracing the original wooden form block pattern on new pieces of plywood and cutting out the patterns just inside the line. This left a similarly shaped form block that was slightly smaller than the original. The aluminum sheet was clamped between the two wooden blocks and the flanges bent over the edges of the larger form block. A rubber mallet was used to ensure that the flanges tightly conformed to the wooden form, and that the radius of each edge was uniform.

The completed sections were then aligned using the original drawings, and holes were drilled into the aluminum to accommodate rivets that would eventually hold the shells together. With the holes drilled, the aluminum shells were stripped off the wooden form blocks, leaving a newly formed metal shape.

⇒ The aluminum flanges are clamped together to create a solid form from several delicate aluminum pieces.

⇒ Thicker plates can be added and riveted around the axles and other high-stress positions. These provide more stability and firmness to the structure.

⇊ After hammering the flanges down around the wooden form block, the aluminum is left with square edges of a small radius when the form is removed.

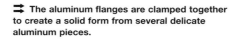

⇈ Cleco fasteners temporarily hold the aluminum members together before they are riveted. These temporary fasteners are much easier to remove than rivets if modifications are made to the design.

The ribs and bulkheads of the robot were assembled first, followed by the skins. Stiffening plates were added to the structures to create mounting locations for other components. Cleco fasteners, temporary fittings that were later replaced with rivets, were used to hold the skins in place. The Cleco fasteners—spring-loaded plungers that allow for temporary fastening of the skins—could be removed if changes were needed.

Once components were finalized and the fit was satisfactory, rivets replaced the Cleco fasteners. This construction method provided the flexibility to alter the robot's design up to the point when

rivets were inserted. Because the Cleco fasteners could be easily inserted and removed, they were favored as an initial means to test the fit. The final product was an aesthetically pleasing frame that was both lightweight and rigid.

Monocoque design provided Team 1501 with a sophisticated robot frame created using simple tools. Recognizing the value of this construction method and acknowledging that it is not commonly used to construct robot bases, Team 1501 promoted this construction method to other teams.

The team developed a step-by-step set of instructions to teach other teams how to use the building technique.

Instructions and an instructional video demonstrating each step in the construction process, were posted on the team's website and shared with other teams.

By using a classic construction method, Team 1501 was able to create a modern-looking and fully functional robot base. The lightweight base was created using hand tools, and had the strength needed for FIRST competitions. By sharing their work with others, Team 1501 made it possible for other teams will be as successful as they were with monocoque construction.

↓↓ The finished product is a modern-looking robot, built with a classic construction method that can serve as an inspiration to other FIRST teams.

# SECTION 02

The General Motors Industrial Design Award celebrates form and function in an efficiently designed machine that effectively achieves the game challenge. This award is based on the judges' review of the team during the competition, looking at factors such as the robot's scoring ability, autonomous program effectiveness, and defensive characteristics. Winners of this award achieved superior on-field performance with well-designed robots and skilled robot operators.

General Motors is a Founding Sponsor of FIRST, and has been involved intensively since its inception, generously supporting events and dozens of teams from both its Michigan base and throughout the country. Senior executives have served on FIRST's board of directors. A hallmark of General Motors' involvement has been the strong mentoring culture it has nourished through the teams it supports. Professional engineers and technical staff from General Motors serve as mentors and role models for high school students, who in turn mentor middle school students to bring them the excitement and inspiration of science and technology.

# General Motors Industrial Design Award

# RHODE WARRIORS: A FAMILIAR ROAD TO OUTSTANDING DESIGN

# THINKING AHEAD WITH PROTOTYPING

Great machines begin with great design. Whether complex or simple, if the robot reliably performs the duty it was designed for, it will be a competitive machine. One method to improve the chances a design will work is to prototype all functions during the design and construction process. Through prototyping, all systems can be tested and improved. This process saves time and avoids potential flaws progressing into the final design.

Team 121, the Rhode Warriors—from Newport County, Rhode Island, and sponsored by U.S. Navy Undersea Warfare Center, Raytheon Corporation and the University of Rhode Island—relied on prototyping to test ideas and to develop concepts into high-performing robot functions.

## ⇄ Moving from Brainstorming to Modular Design

The design process began with a brainstorming session to generate initial ideas. Suggestions were collected, reviewed, and analyzed to create an initial design for the robot. With a plan for the robot established, the team divided into subteams to design each of the robot's subsystems.

In addition to being assigned a task for the robot to achieve, each subteam was allocated a particular space inside

⏬ **A preliminary white-board drawing shows the layout of the robot and where different mechanisms should be placed.**

⏬ **An AutoCAD drawing is used to build the upper aluminum frame, called the silo, which can be bolted on to the chassis.**

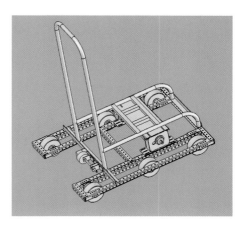

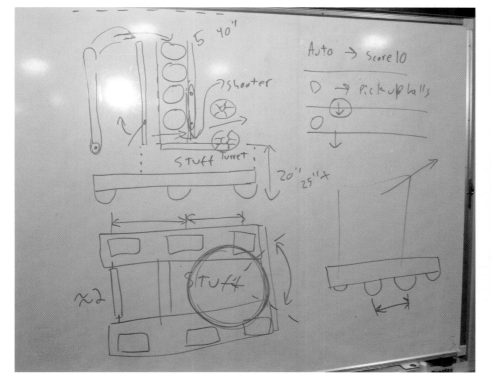

the robot in accordance with a preliminary layout of the machine. The subsystem had to fit in the assigned envelope and operate effectively with other robot systems. The integration of these subsystems early in the design process avoided complications and ensured the final components would be compatible with each other.

The robot was constructed in modular sections that were built independently and later assembled. The two main structural components were the chassis and the section the team referred to as the "silo," which was bolted to the chassis. The silo was the robot frame and upper structure that contained the robot's scoring system. The ball manipulation, storage, and shooting systems were designed to fit within this frame.

Because these two main sections were separate, construction on each occurred simultaneously. In addition, each system could be maintained separately. Each component was prototyped to test its effectiveness, with improvements made until the system was designed as intended. Following testing, the prototypes progressed to final designs and were assembled together.

⇉ The chassis and the "silo" were designed and built separately. When integrated, the parts comprised the final robot.

The robot chassis housed two components: the drive system and the ball collector. The team first constructed a wooden prototype of the frame and wheels to test its clearance with the ramp. The team wanted to use the chassis provided in the FIRST Kit of Parts with a six-wheeled drive system. Satisfied with the prototype's performance, the team proceeded to modify the chassis to accommodate the small wheels used in its design.

To reduce costs and save time, the gearbox transmissions, wheel hubs, axles, and motor mounts supplied in the FIRST Kit of Parts were also used on Team 121's robot. This approach proved to be more beneficial than a custom design because of the ease with which the components integrated and their high degree of reliability. In addition, by electing to adopt a proven transmission system, more time was available to work on the design and integration of other subsystems.

The robot's electronics were mounted on a single removable piece of polycarbonate plate. This panel attached to the robot chassis from the bottom and could be easily dropped out for alterations or repairs. This orientation added several inches of vertical space within the chassis—a valuable commodity in FIRST robots. The battery was mounted at an angle to allow for easier access.

⬇⬇ The chassis frame provided in the kit of parts presents Team 121 with an easy-to-construct base for the rest of the robot components.

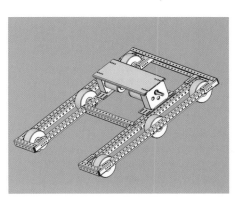

The silo housed a rotating turret, the ball storage device, a ball lift, and a shooter mechanism. Each system was first developed as a prototype from the same aluminum tube used on the final design. By using the same material, the prototypes became templates for determining the size and location of each subsystem.

The entire silo rotated on a circular disk supported by ball bearings at its base. A rubber wheel, powered directly from a window motor supplied in the kit of parts, controlled the rotation. This wheel was mounted on a spring-loaded arm to maintain positive contact with the silo. All systems on the silo were driven directly by motors provided in the kit of parts, thereby eliminating the need to manufacture parts or incorporate speed-reducing transmission systems.

⬆⬆ A polycarbonate sheet at the bottom of the robot can be easily removed to repair the electronic components mounted on it.

⬇⬇ The battery, mounted at an angle in the back of the robot, can be easily removed or replaced by a new battery for the next match.

# ⇄ Optimal Designs From System Comparisons

A series of wheel prototypes was constructed to determine the best way to launch the balls into the upper goal. The best design consisted of two wheels pressed onto an aluminum hub. The wheels were powered with the largest motor from the kit of parts that was directly coupled to the wheels.

Testing revealed that the lightweight wheels lost a significant amount of rotational momentum every time a ball was shot. The team considered using a heavier wheel to maintain inertia but this option was not selected. The heavier

⇄ A dual-vertical-wheel prototype is constructed to test one of the suggested methods of launching foam balls into the high goal.

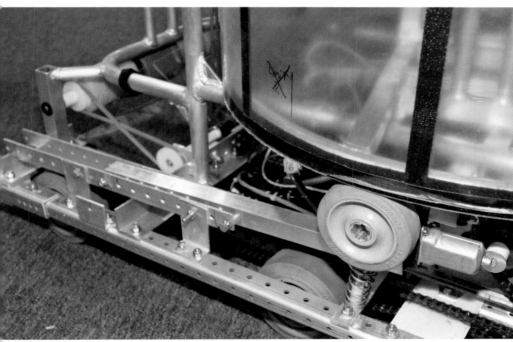

⇈ The rubber wheel, mounted on a spring-loaded arm, can be seen here. As a motor turns this wheel, it causes the silo to rotate.

⇄ Another wooden shooter prototype is constructed, using two wheels mounted on a single axle. This prototype is the basis of the final shooter design.

wheel would have resulted in a significant weight addition to the top of the robot and have decreased the robot's stability.

A backstop constructed from bent aluminum tubing was designed to pinch each ball and create backspin when the ball passed through the launch mechanism. The team considered adding a tilting mechanism to manipulate shooting distances but decided against this option as well, because of weight considerations. Instead, it was determined that range would be varied by altering the motor speed on the shooting wheels.

A camera to track the target light was mounted on top of the turret. Since the turret was capable of rotating 180 degrees, the pan motor supplied with the camera mounting system was not needed. The tilt of the camera was controlled by a servo motor.

When pointing toward the goal, the tilt angle of the camera was measured using the output of the tilt servo motor. The measured angle was compared to a predefined database of the target range for each measured angle. With this data point, the speed of the shooting wheels was automatically adjusted to launch the ball the required distance. The flexibility to change shooting speed provided the capability to shoot from any distance to the goal and avoided adding mechanical components to the shooting system.

⬇⬇ **A CMU2 Camera is mounted close to the shooting wheels at the top of the silo. When the turret rotates, the camera pans to find the high goal.**

# ⇄ A Robot Silo: Ball Storage and Transport

Situated within the silo was the ball storage device—a helical shape to efficiently accommodate a large number of balls. Prototypes of the storage helix began with ramps that could be varied to evaluate ball storage and movement. The helix design was based on its ability to hold the most balls in a limited volume.

Balls were collected off the playing field floor with a roller, driven by a figure-eight cord. These balls were transported to the top of the robot via an outer, cord-driven lift. They were deposited at the top of the helix where they then rolled down the ramps in a controlled linear order. When the ball sreached the bottom, they were again lifted to the top of the robot through the center of the helix to be fed into the shooting mechanism.

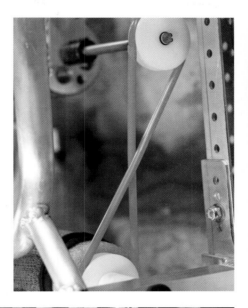

⬇⬇ A pulley-powered cord provides rotation to the roller, which pulls the foam balls into the front of the robot as it rotates.

⬇⬇ Team 121's robot is skilled at collecting balls off the playing field floor and conveying them to the top of the robot, where they can be shot at the goal. Here, the robot is seen practicing its shots on goal.

⬆⬆ The balls are lifted to the shooting wheels through the center of the storage helix. They are conveyed between polycord belts and aluminum tubing.

⇄ The circular ramps of the helical storage bin can store a large amount of balls while keeping them organized in a linear fashion.

The systems to collect and lift balls were similarly constructed. Motors located near the robot base provided power to each system. A lower drive pulley rotated three poly-cord belts that were wrapped over an idler pulley mounted high in the silo. As the pulleys rotated, the belts were driven and the balls traveled along them.

Each pulley originally had individual grooves to hold each poly-cord. Through testing, it was found that when the balls approached the idler pulley the belts would ride out of their respective grooves and cause the system to jam. Because the lift system was already designed as an integral part of the silo and its subcomponents, the layout and volume of the lift could not be changed. Any improvements were restricted to the space envelope of the original lift system.

Through brainstorming and additional testing, Team 121 discovered that the individual grooves on the idler pulleys could be eliminated. In this configuration, the poly-cords would run together in a single, wide groove. The belts did not stretch enough to ride off the new wide pulley, and, as result, the system stayed intact.

To accommodate for the decrease in shooting wheel speed as each ball was propelled, the speed of the shooter's ball lift to the shooter was regulated. With the correct ball lift speed, the shooting wheel could return to its full speed and regain momentum between shots. Although slowing the rate of balls limited rapid firing, the robot made up for that with its accuracy.

⬇⬇ Removing the grooves in the pulleys makes the polycord belts much less likely to detach from the system as they ride in a single, wide groove.

⇋ The polycord belts of the conveyor system originally fit into grooves in the controlling pulleys, seen here at the base of the system that transferred balls to the shooting wheels.

‼ The final product of Team 121's six-week build period is a modular robot with easily accessible components, which retains a straightforward and efficient design.

## ⇄ Realizing the Benefits of Prototyping

Team 121 used extensive prototyping throughout the robot design process. This approach provided the team with physical models of each component to study, manipulate, and modify before the final components were built. The prototype systems gave the team confidence that each mechanism worked and fit as designed within the robot.

The prototype process also helped Team 121 save time and money—rare and precious commodities in the world of FIRST. The team's experience working with prototypes refined their ability to troubleshoot systems and resolve problems. This skill was beneficial when the team faced design challenges such as with the pulleys. Instead of abandoning the system, they drilled into the issue to find the best solution. Team 121's planning, prototyping process, and simple design philosophy were important waypoints in the Rhode Warriors' journey that produced an excellent, high-performing robot.

# PLAN, DESIGN, IMPROVE: THREE STEPS FOR HIGH PERFORMANCE

## AS EASY AS COUNTING 1, 2, 3

The members of the Hamtramck (Michigan) High School FIRST Team 123 understood that creating a winning robot could be as easy as counting to three. For this team, the numbers 1, 2, and 3 represented important steps in the iterative design process: plan, evaluate, and improve. As practiced by engineers from the team's sponsoring companies (General Motors, Cadence Innovation, Coffey Machining Services, ITT Tech, and Ford), the team realized the need for careful analysis before beginning a design, validating plans before construction, and improving the physical design based on real-world tests. These steps made building a winning robot almost as easy as 1, 2, 3.

# ⇄ Analysis: The First Step

Team 123 started the design process by creating a computer model of the field and then filling that field with the 80 game balls allowed on the field floor. The crowded field convinced the team that an effective robot would need to collect balls from both sides. Further, the tight location of balls suggested that a highly maneuverable robot would have a strong advantage. Finally, the team recognized that collecting a good number of the 80 balls and depositing them in the lower goal, combined with the ability to play aggressive defense, were attributes that could result in a highly competitive robot.

The team decided early in the design phase not to shoot balls, but rather to collect and deposit them into the lower goal. This simplified their overall design and allowed them to concentrate resources on perfecting this capability of their robot. The team decided that keeping the weight low and in the center of the robot would result in a stable robot that would resist tipping during collisions or while climbing the ramp.

A common system was designed to pick up balls on both sides of the robot. This system consisted of a series of timing belts driven off a common shaft. These conveyor systems could be operated in reverse to drive balls out of either side of the robot. A small arch, designed to allow the balls to flow from one side of the robot to the other, separated the two ball-handling systems.

The drive system powered all four wheels, with high friction wheels used in the rear of the robot and omni-wheels used in the front of the robot for increased maneuverability. Poly-carbonate was selected as structural material to allow quick and easy visual inspection of the robot.

⇈ The robot's view of the field illustrates the obstacles created by a large number of balls on the field. A quick and effective ball retrieval system is the only hope for maneuvering around such a congested field.

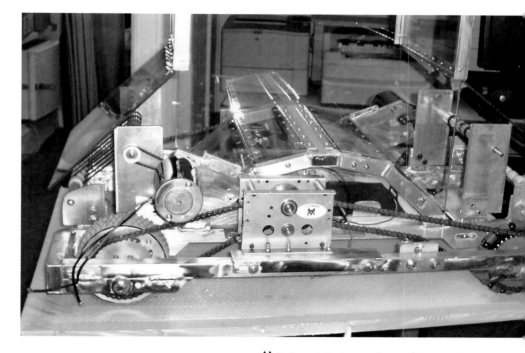

⇈ Rollers at the front and rear of the machine provide a means to pick up and deliver balls from each side of the robot.

## ⇄ Validation:
## The Second Step

Every aspect of the robot was validated before construction. Computer models of the field and robot were important tools for Team 123 during this second step of their design process. By closely examining the robot computer model, the team understood and avoided problems before constructing the actual robot.

The team desired a low center of gravity, and this required a low robot frame. The team also wanted a robot that could climb the ramp at the end of the match, and this performance was most easily accomplished with a high robot frame. To resolve this dilemma, the optimal clearance height between the robot frame and the field floor was determined using computer analysis.

The team had already created the computer model of the playing field elements, and this model was used in conjunction with a model of a preliminary robot base to check the clearances between the robot and the field. Variables that had to be considered were diameters for each set of wheels, robot frame length, chain paths, and the mounting location of each wheel along the frame.

⬇⬇ **A "kicker" lifts balls from the rear of the robot to the front. The arch at the front of the robot allows the balls to roll out of the robot and into the lower goal.**

⬇⬇ **Computer modeling assures that correct choices are made for the wheels and their spacing on the frame. By minimizing clearance, the overall stability of the robot is enhanced.**

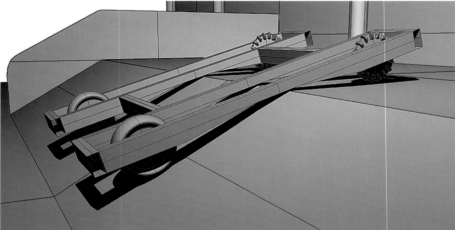

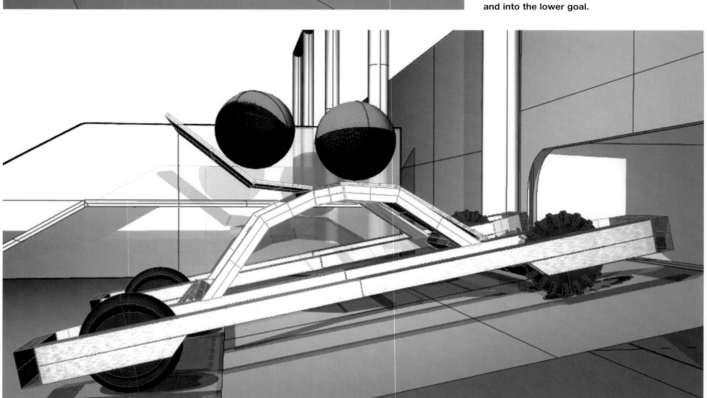

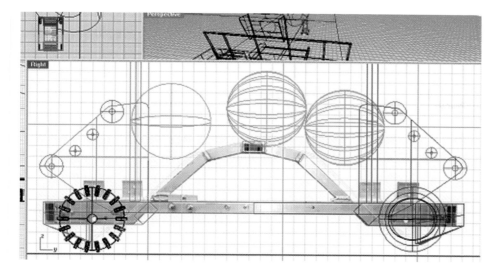

A comparison of the constructed frame with the original design is accomplished in this composite image. A photograph of the frame is superimposed on the computer model to confirm the correct shape has been constructed.

The computer models made it easy to examine various combinations of components, to ensure the robot would perform as expected.

Once the frame geometry was determined, the basic frame was augmented with an arch in the middle of the robot. This model was placed on the ramp for the lower goal and balls were added to the model. A review of this composite model indicated that the trailing edge of the arch would have to be lifted to get balls to roll off the robot into the lower goal. The team called this moveable component the "kicker" as it tended to kick the balls out of the robot. Here, too, the clearances between the robot, balls, and goal could be immediately determined from accurate computer models.

The team routinely used the computer model to validate the manufacturing process. For example, they photographed a side view of the manufactured frame and imported that image into the CAD model of the robot. By overlaying the photo image on the CAD model, the accuracy of the manufactured component was verified.

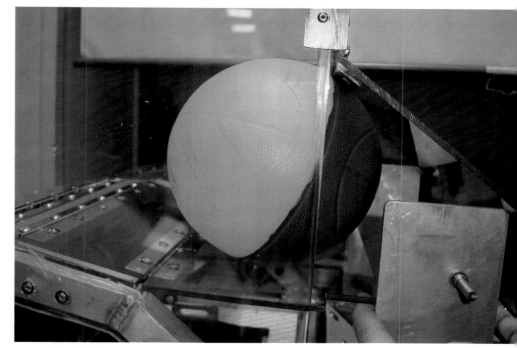

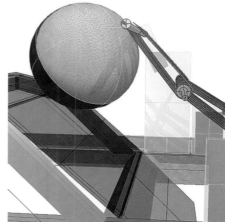

⇄ ⇈ Prototype testing indicates that a pinch point exists between the ball and the upper and lower robot surfaces once the ball exits the rollers. Computer modeling illustrates that the problem can be eliminated with a second-stage of conveyor belts to transport the balls.

# ⇄ Testing and Improvements: The Third Step

The design process is often referred to as being iterative, where changes and improvements are made at each step. Team 123 found great value in this aspect of the process. Rigorous testing and evaluation revealed a number of improvements that increased the robot's ability to function.

The team's ability to amend the computer model to redesign components was of great benefit during this phase of the project. Early testing revealed that a single set of belts would only lift the balls to the bottom edge of the ball hopper. This created a pinch point between the ball and the upper plate

⇈ A second-stage conveyor system, driven by the first stage, lifts balls into the ball reservoir and prevents jamming.

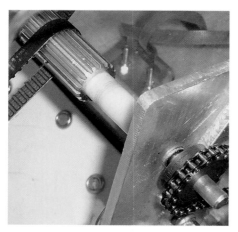

⇄ Originally, a timing belt is used to power the ball-retrieval system, but a jam in the conveyor strips the belt. A chain-and-sprocket drive replace the belt drive and improve the robot's reliability.

when the kicker was activated. To eliminate this problem, the location for a second set of belts was determined using the computer model. The second set of belts was put in a location that would not interfere with the kicker lift mechanism. Each side of the robot was then modified with the addition of a second-stage ball lift driven off the same shaft as the first stage conveyor.

The conveyor systems were initially powered using a timing belt between the motor and the shaft that supported the belts. In one practice session, a ball jammed in the conveyor. The motor continued to operate, thereby shredding the drive belt that was prevented from

rotating because of the jam. A simple change—replacing the drive belt system with a chain drive system—eliminated this problem and proved to be a more reliable connection.

Testing also revealed a tendency for the balls in the original design to jam when they were loaded in the robot. The straight sides of the ball hopper did not allow the balls to flow out of the robot. To correct this problem, the side panels of the polycarbonate hopper were bent, each at a different height from the ground to prevent the balls from stacking when emptied into the goal. In addition, this change provided more storage capacity.

⇄↓↓ The original straight sides in the hopper prompts balls to jam against one another. An alternative design with sloping sides eliminates this problem and expands the robot's capacity to carry balls.

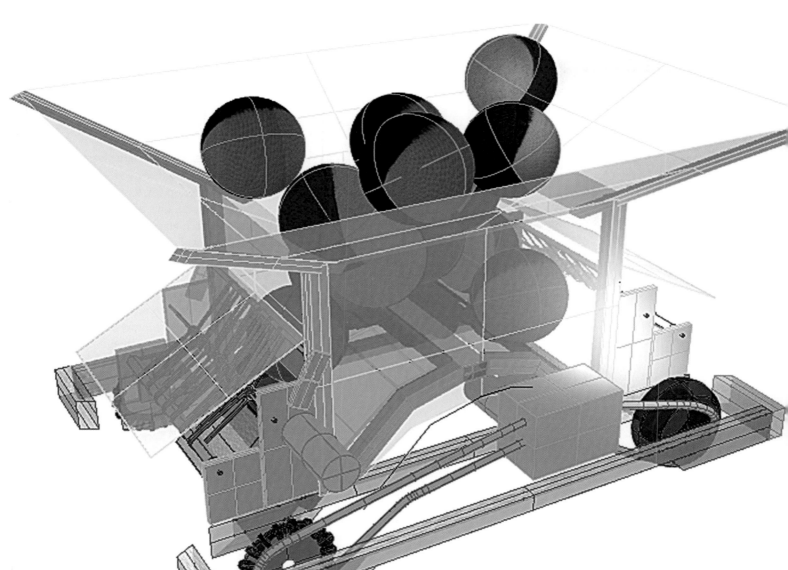

⬇⬆ The extended piston keeps the "kicker" closed. When the piston is retracted, the "kicker" raises and lifts balls over the robot's mid-section.

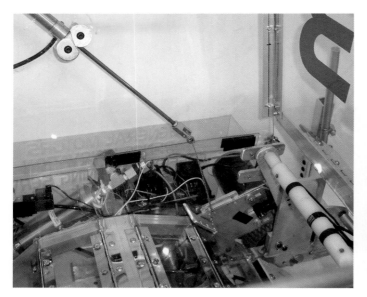

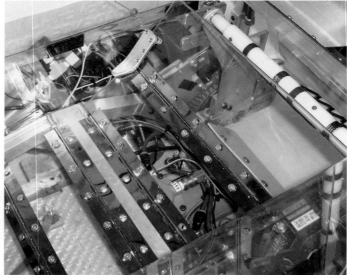

⬆⬆ By mounting two omni-wheels side by side—offset approximately 20 degrees—one of the rollers would always be in contact with the group. Single-wheel designs tended to fail in the region between rollers in certain conditions when the hub would make contact with the playing field.

The lift system for the kicker was also evolutionary. Early plans called for a pneumatic piston located below the kicker plate to raise the kicker and push balls out of the robot. Construction constraints prevented the piston being located beneath the plate. Instead, a piston was mounted above the plate on the side of the hopper. A cord connected the piston end to the kicker, and when the piston was retracted the kicker plate was raised.

The omni-wheels also benefited from testing. The plastic bodies of the omni-wheels did not hold up well against the aluminum ramps on the playing field. To eliminate this problem, two omni-wheels were used on each side, with the rollers offset so as to always have a rolling surface in contact with the playing field. This change greatly extended the life of each omni-wheel.

## ⇶ The Final Robot: Putting It All Together

Comparing Team 123's robot before and after competition revealed additional improvements. Vertical bars were added to each corner of the robot to prevent it from entering the lower goals—a prohibited action that brought heavy penalties during match play. Also, a front bumper was added to the robot to protect the conveyor lift. In addition, the polycarbonate shield in front of the robot was enlarged to provide room for any balls that forced their way through the conveyor belts. These changes were all made during the course of a FIRST Robotics Competition, proving that it is never too late to improve a design.

Team 123 executed the three important steps of the design process. The Team's designers analyzed the requirements for the competition, used sophisticated modeling techniques to validate their design, and continually improved their design through rigorous testing. And throughout this process they documented their work, thereby creating a design and decision log to fully to explain its systematic approach to building a competitive robot.

↑↑ A before-and-after comparison highlights the changes made to the robot after participating in a regional FIRST Robotics Competition. Features were added to protect the robot from penalties and damage during play.

# ELECTRONICS, MATHEMATICS, AND ROGRAMMING —OH, MY!

## COMBINING MECHANISMS AND CONTROL OPERATIONS TO BUILD A BETTER ROBOT

The design process and competition is not limited to the mechanics of the machine but also includes the means by which the robot is controlled. For the 2006 FIRST Robotics Competition, many robots were created with a drive system, a ball scoring mechanism, and a ball collection device. The most successful of these machines incorporated ingenious mechanisms and sophisticated controls to accomplish the game's challenges.

Team 190, from Worcester, Massachusetts, stood out among the competition for the ingenuity and cleverness manifested in their robot and controls. Mathematical analysis and an extensive variety of sensors were used to monitor, guide, and predict system performance. Their work served as a case study on integrating a control system with a well-engineered machine to create a superior robot.

## ⇥ Modeling and Matrices

When the FIRST 2006 game was announced, Team 190 anxiously watched the details over an Internet broadcast at Worcester Polytechnic Institute in Worcester, Massachusetts. Ideas were generated even before the kick-off presentation was over. A prioritized list of possible strategies soon emerged, and the team began presenting its ideas and sketches for a robot that would dominate the competition.

With so many ideas, the team's designers needed a way to narrow them into one design. They accomplished this through use of an engineering decision matrix that compared design attributes for different robots. Categories covered by the matrix included the ability to score on the ramp, power consumption, drivability, low design risk, attractiveness, speed, agility, ease of programming, robustness, and, of course, weight and were weighted for their importance relative to all other categories.

Each robot design idea was presented to the team. Team members evaluated each design for every category listed in the decision matrix. The scores were collected and assembled to determine which robot would be constructed. The design chosen

⇆ A well-designed control system assists in the management of the components that make up Team 190's robot.

⇊ ⇥ Sketches were a large part of the initial brainstorming process, when team members expressed their design ideas.

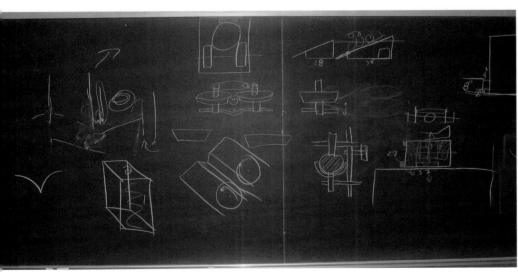

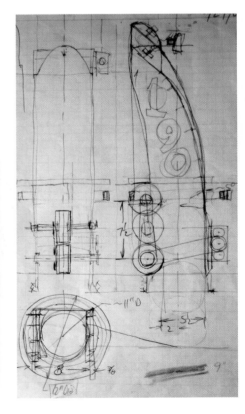

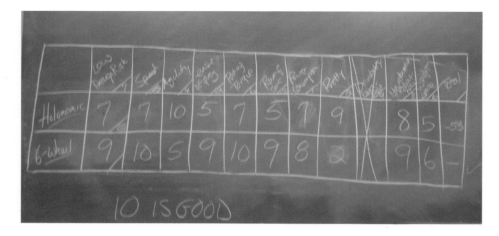

The matrix image contains a hand-drawn chalkboard table:

| | Low Weight Rob | Speed | Agility | Cool Rating | Turning Torque | Pushing | Power Consumption | Pretty | Durability | Development Learning Curve | Difficulty to Build | Total |
|---|---|---|---|---|---|---|---|---|---|---|---|---|
| Holonomic | 7 | 7 | 10 | 5 | 7 | 5 | 7 | 9 | | | 8 | 5 | -55 |
| 6-wheel | 9 | 10 | 5 | 9 | 10 | 9 | 8 | 2 | | | 9 | 6 | - |

10 IS GOOD

⬆⬆ An engineering matrix is constructed to weigh and compare the different aspects of components. Seen here is a matrix comparing a holonomic drive with a six-wheel drive.

through this analysis beat out the other designs by a margin of 15 percent. With a design chosen, sketches that had been drawn by hand were reviewed, with more detail added as a clearer definition of the robot came into being. The hand sketches were next converted into three-dimensional computer models using PTC Pro/Engineer and Autodesk Inventor CAD software. Each component on the robot was first modeled separately using CAD. A robot was later created with CAD by assembling the virtual parts into a composite structure.

The computer models allowed the team to visualize the entire robot before any parts were manufactured. This allowed them to examine how the components fit together and if the size requirements were being met. This level of planning avoided problems, saved time, and conserved resources by manufacturing only the correct parts.

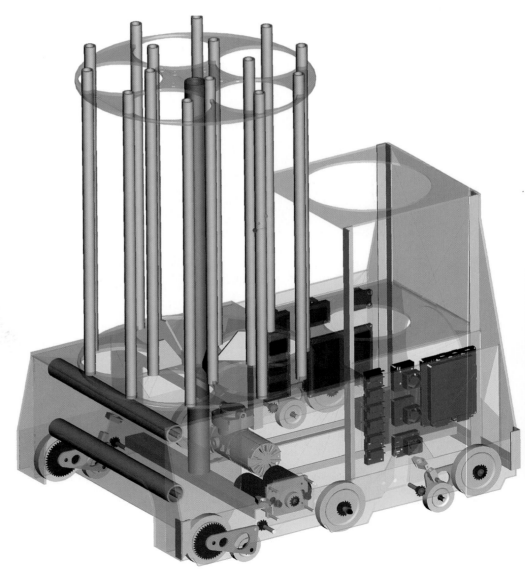

➡ Computer models provide early insight of the robot design. Each component is created with sophisticated, three-dimensional modeling software and then assembled to build the complete robot in a virtual environment.

# ⇄ Parabolic Pondering

The robot design chosen by Team 190 included a mechanism to score in the upper goals. The team realized the need for a low center of gravity and, with this in mind, designed the mechanism to launch balls from low on the robot. The balls followed a curved path up the robot to a deflector plate at the top of the robot. With an exit height of 5 feet (1.5 m), the shooting system was designed to make it hard for other teams to block shots made on the goal.

Possible arrangements of the shooting system were prototyped using a variety of wheel combinations. The final design consisted of two wheels, stacked vertically. Analysis determined the wheel speed needed to launch the balls. The wheels were geared to spin at different speeds to launch the balls with a desirable amount of spin.

The team determined the lower wheel's speed should be 4,000 revolutions per minute (rpm), and the upper wheel, 7,000 rpm, to boost the exit velocity to the maximum limit allowed of 39.4 feet (12 meters) per second. With this combination of speeds, each wheel imparted approximately half of the power needed to accelerate the ball. The resulting backspin flattened the ball's trajectory and made aiming more reliable.

An optical encoder monitored the wheel speed, and a loop in the control software regulated the speed if it strayed from its set value. This ensured accuracy and imparted control on the motors that powered the shooting mechanism.

The parabolic shape of the launcher was selected based on the same principles that determine the shape of common car headlights. Headlights contain parabolic mirrors that surround the bulb, that is in the center of the headlight. Light waves that leave the bulb bounce off the mirrors and are deflected straight out from the headlight.

This concept was modified for the shooting system. The balls represented the light waves, and the deflector represented the mirrors. With the correct adjustment, the balls left the robot with a predictable trajectory, regardless of where they bounced off the deflector. This system allowed the robot to reliably score at distances between 6 and 21 feet (1.8 and 6.4 m).

↑↑ With the launching mechanism low on the robot and a large deflector, the center of gravity was minimized.

⇄ This view shows the lower of two shooting wheels, mounted vertically. The foam balls are directed between these two wheels to be shot into the high goal.

FRIENDS OF STEVE KATZ

↓↓ The parabolic shape of the deflector is carefully calculated to provide a constant trajectory for the foam balls that are launched by the shooting wheels.

## ⇄ Focus on the Target

To rotate the camera towards the goal without necessarily moving the chassis, Team 190 mounted the camera and shooter on a turret with a 540-degree range of motion. The turret was constructed from a polycarbonate sheet that had an 8-inch (20.3 cm) hole bored through the center to move balls through. A potentiometer was used to measure the turret's rotational position.

A two-stage spooling system was designed to control the wires that supplied power to the turret components. The spools of wire rotated with the turret and let out wire as needed when the turret rotated. The wires moved opposite each other on either side so that as one set of wires was taken up the other was released. Excess wire ran around pulleys mounted in tracks at the sides of the robot. A shock cord connected to each pulley kept constant tension on the wires to prevent entanglement.

The camera was mounted at a fixed angle and the turret automatically adjusted the shooter to focus the camera on the goal. While the turret was locked on the target, gear tooth sensors on the drive motors and a gyro sent data to the robot controller, which computed the speed and direction of the robot.

Using this data, a program instructed the turret to lead or lag, to compensate for the speed and direction of the robot. When the robot was bumped, for example, the gyro data alerted the turret to compensate for the sudden movement to keep the camera and turret centered on the goal. With this sophisticated goal tracking, the robot could simultaneously complete other tasks while scoring.

↓↓ The polycarbonate turret ring provides the robot with a ball-shooting system with a large range of rotation. The turret could rotate a total of 540 degrees, independent of the chassis.

↑↑ The aiming system constantly adjusts the shooting mechanism independently of the robot's position, allowing foam balls to be accurately launched into the high goal.

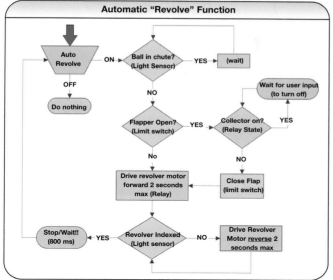

## Automatic "Revolve" Function

Auto Revolve — ON → Ball in chute? (Light Sensor) — YES → (wait)

OFF → Do nothing

Ball in chute? NO → Flapper Open? (Limit switch)

Flapper Open? YES → Collector on? (Relay State) — YES → Wait for user input (to turn off)

Collector on? NO → Close Flap (limit switch)

Flapper Open? No → Drive revolver motor forward 2 seconds max (Relay)

Drive revolver motor forward 2 seconds max (Relay) → Revolver Indexed (Light sensor)

Revolver Indexed YES → Stop/Wait!! (800 ms)

Revolver Indexed NO → Drive Revolver Motor reverse 2 seconds max

User Input
Sensor Input

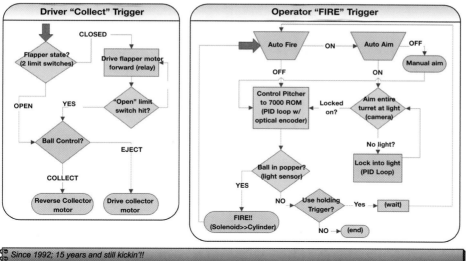

## Driver "Collect" Trigger

Flapper state? (2 limit switches) — CLOSED → Drive flapper motor forward (relay)

Drive flapper motor forward (relay) → "Open" limit switch hit?

"Open" limit switch hit? YES → Ball Control?

Flapper state? OPEN → Ball Control?

Ball Control? COLLECT → Reverse Collector motor

Ball Control? EJECT → Drive collector motor

## Operator "FIRE" Trigger

Auto Fire — ON → Auto Aim — OFF → Manual aim

Auto Aim ON → Aim entire turret at light (camera)

Auto Fire OFF → Control Pitcher to 7000 ROM (PID loop w/ optical encoder)

Control Pitcher — Locked on? → Aim entire turret at light (camera)

Aim entire turret at light (camera) → No light? → Lock into light (PID Loop)

Control Pitcher to 7000 ROM → Ball in popper? (light sensor)

Ball in popper? YES → FIRE!! (Solenoid>>Cylinder)

Ball in popper? NO → Use holding Trigger? — Yes → (wait)

Use holding Trigger? NO → (end)

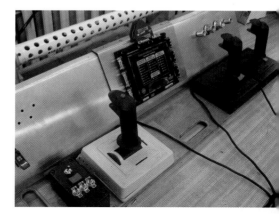

The control board is set up for a driver and an operator to direct the robot. The three main operator-controlled automatic functions are seen mapped out in a flow chart (left).

## ⇄ Controlling the Robot

Two team members managed the robot during competition. The driver controlled the steering and the ball collector. The operator was in charge of the shooting and revolving mechanisms. There were three main automatic functions the operator activated to coordinate robot actions: auto-aim, auto-fire, and auto-revolve.

The auto-aim function initiated a sequence of action to automatically locate and lock on the target. Auto-fire was activated with a joystick trigger to continuously fire balls at the goal. The auto-revolve function rotated the barrel-styled ball collector to the next chamber that contained balls. The barrel rotated until it reached a position where an optical sensor detected the presence of balls in the chamber.

The auto-fire condition could only be activated if certain conditions were met. The camera had to be locked on the goal, the shooter wheels needed to be rotating at the correct speed, and a ball had to be in firing position. The auto-fire function could be disabled, thereby allowing the operator to manually fire balls at the goal.

Other control features enhanced the robot's performance. A switch was used to control the shooting and tracking functions, turning them on or off depending on whether the robot was playing offense or defense. Another switch disabled the robot's tracking system if the robot turret reached its

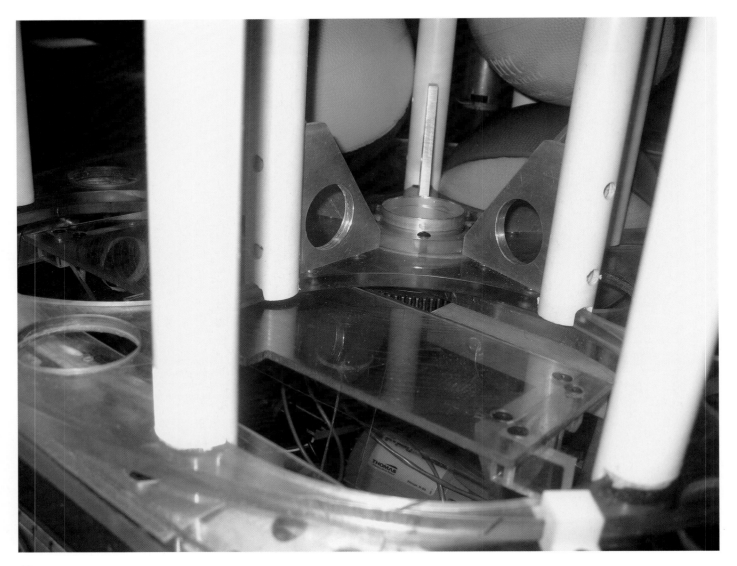

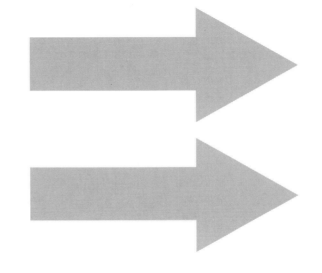

**↑↑** The balls are stored in a cylindrical, revolving system of chambers. This system automatically senses the position of foam balls and sends them to the shooting wheels. Seen here is a view inside one of the chambers.

rotational limit. A "spinup" button brought the shooting motors up to launch speed when the robot was preparing to score. There was also a reset switch, which reset the code but did not reinitialize the camera or gyro.

To monitor the array of sensors and robot conditions, Team 190 mounted lights on the control board. These lights relayed the ball-collector position, camera status, and turret angle, and signaled if the target was to the left or right of the robot.

## ⇄ Models, Mechanisms, and Monitoring Robot Performance

The ability to drive, score, and defend are great assets to any robot. Maintaining precise control over these activities is another story. Team 190 used a structured approach to gather ideas and select the best option. Its careful planning was exemplified in the accurate virtual model of the entire robot that was produced to verify the design.

Team 190 demonstrated its engineering expertise by building a sophisticated robot and designing a control system to maximize performance. Aided by a large number of sensors and team-written computer code, Team 190 commanded the robot's movement, storage capability, aim, and scoring ability. Automatic functions increased the robot's efficiency, and the ability to manually control the individual mechanisms averted unpredicted problems. In the end, Team 190 combined form and function to create a powerful and victorious machine.

⇅ Extensive programming and implementation of a complex control system supply Team 190 with great command over their robot.

# ONE TRIUMPHANT TURRET

## CAMERA AIMING AND RAPID SHOOTING

A rotating, camera-controlled turret was the heart of the Team 237 robot, built by students at Siemons-Watertown High School in Watertown, Connecticut. The turret—topped with a ball launcher and integrated with a ball retrieval and conveyor transport—was the choice mechanism designed during the team's initial brainstorming session. The camera was mounted on the turret to continuously track the light and aim the shooting system at the goal. Whether being defended against by other robots, or driving around picking up balls, the turret-mounted, camera-aimed shooter was always ready to launch a ball and score.

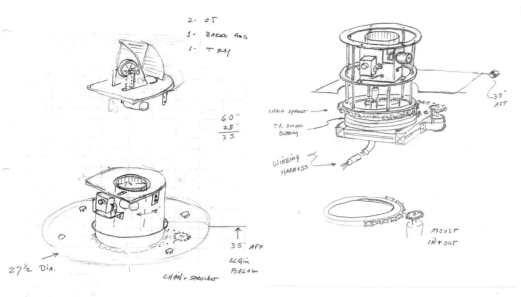

The design process begins with thoughts and ideas, often recorded as sketches. Significant amounts of detail in the sketches add clarity to the design and serve as a visual template for refinement.

## ⇄ Designing a Terrific Turret

Team 237 foresaw that a winning robot would need to be able to shoot balls while moving, thereby remaining a scoring threat when being jostled by other robots. Based on these factors, a turret was preferred over a fixed shooting mechanism that could only be aimed by steering the robot. This design strategy paid off handsomely as the team was able to create the turret and create a high-scoring robot.

The team's goal was to build a turret that served as a foundation for the robot's tracking and launch systems. Hand-drawn sketches of the turret and shooting system were a starting point for planning and provided a visual outline on which to base designs. These drawings prompted the team to examine the performance and interdependency of each system on the robot.

The hand sketches were refined using three-dimensional modeling software to create an accurate computer model of the robot. The computer model enabled the team to examine closely each system to ensure it would operate as planned, and avoid any unseen interferences between components. Detailed blueprints of the turret parts were produced using the modeling software, and these plans were used to manufacture the turret components

Originally, the team wanted the turret to rotate 360 degrees to provide it with an unrestricted range of motion. This design would have required sliding electrical contacts to transfer power and control signals between the fixed robot base and the rotating turret.

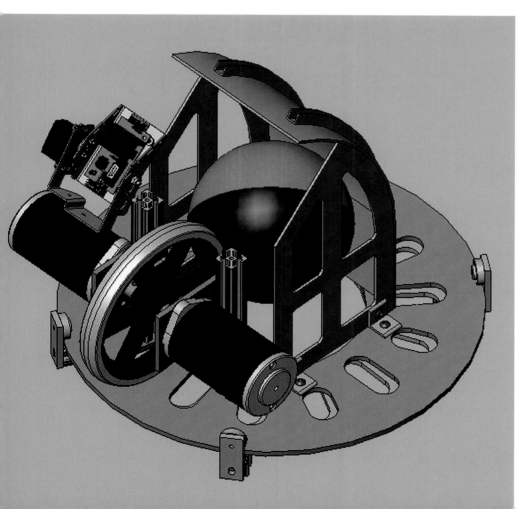

⇄ Team 237 relies on three-dimensional modeling software to create detailed simulations of robot systems. The virtual model is a tool to learn about the systems before construction begins.

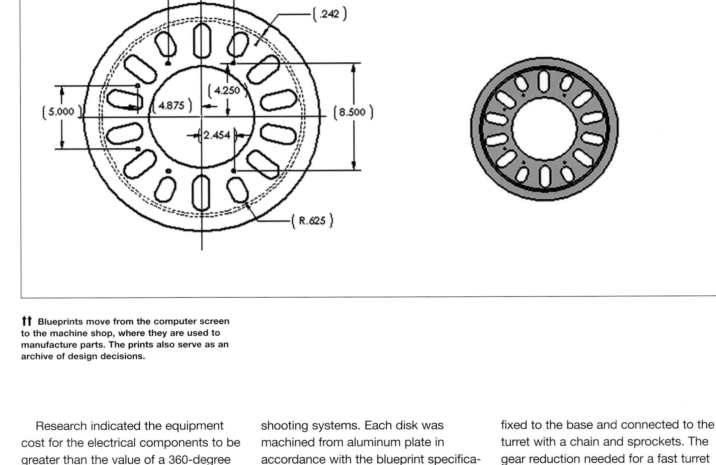

↑↑ Blueprints move from the computer screen to the machine shop, where they are used to manufacture parts. The prints also serve as an archive of design decisions.

Research indicated the equipment cost for the electrical components to be greater than the value of a 360-degree system. Instead, a system was designed that could rotate nearly 360 degrees, using conventional wires to provide power and control between the base and the turret systems.

The turret was designed to be as simple as possible to ease manufacturing and increase reliability. Two aluminum disks were used: one fixed to the robot frame and the other floating on top of the fixed disk. The upper floating disk served as the foundation on which to mount the camera and

shooting systems. Each disk was machined from aluminum plate in accordance with the blueprint specifications. Grooves were machined along the edge of each circular plate to create a track for plastic ball bearings to ride in. These bearings, made from Delrin, transferred the weight of the rotating components on the upper disk to the robot frame. The plastic bearings were selected over steel because of their lighter weight and reduced friction, and because of how easily they allowed the turret to spin.

The first turret-drive system considered by Team 237 consisted of a motor

fixed to the base and connected to the turret with a chain and sprockets. The gear reduction needed for a fast turret required custom-made sprockets that would be heavy and expensive. After careful review, the team decided to drive the turret using a wheel which supported a portion of the upper disk's weight. The wheel was mounted along the disk's outer edge and when it rotated the turret rotated. The friction drive system eliminated weight and was a simpler design than the initial chain and sprocket design. In addition, the fewer number of moving parts increased the system's reliability.

⬇⬇ The turret consists of two disks; the smaller one is the foundation for a larger disk that rides upon it. Four brackets, each consisting of two wheels that act as a clamp, are positioned along the edge of the disks to keep the plates in contact with each other.

⬆⬆ To reduce weight in the upper disk, slots are machined into the plate. The resulting plate retains the needed strength but is greatly reduced in weight.

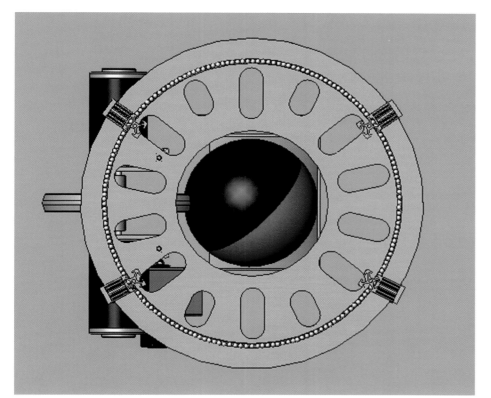

⬆⬆ Grooves are machined into each plate to hold plastic ball bearings that reduce friction between the rotating circular plates. The bearings freely roll in the grooves while supporting the weight of the entire turret.

⬅ A rotating wheel pressing up on the bottom of the upper plate provides the tangential force to rotate the turret. The pneumatic wheel is in firm contact with the plate and is a simple mechanism for rotating the turret.

## ⇉ Turret Control

The turret was designed so it could be automatically controlled by software or directly controlled by the robot operator. The robot's camera, mounted on the turret, provided a signal to the robot's computer program, which processed the signal and made decisions. Team 237 wrote software to rotate the turret in two modes: search mode and track mode. In search mode, the turret rotated until a microswitch reached a physical stop. At that point, the rotation was reversed and the search continued in the opposite direction. Once the target light was detected, the software switched to track mode.

In track mode, the turret rotated to center the light in the camera's frame of view. The speed of the turret's rotation was proportional to the image's distance to the center of the camera frame. As the image approached the center of the camera's frame, the rate of rotation slowed to prevent overshooting the target.

Once the image was centered, the software program continued to monitor the position of the light and constantly adjusted the turret to keep the target in the center of the camera's view. This feature kept the robot's shooting system aimed at the target as the robot moved around the field. Tracking mode proved very valuable during the competition when opposing robots would play defense with constant collisions to deflect shots. Thanks to this system, the shooter remained locked on the target. A manual override capability was included in the software to enable the robot operators to take control of the system at any time.

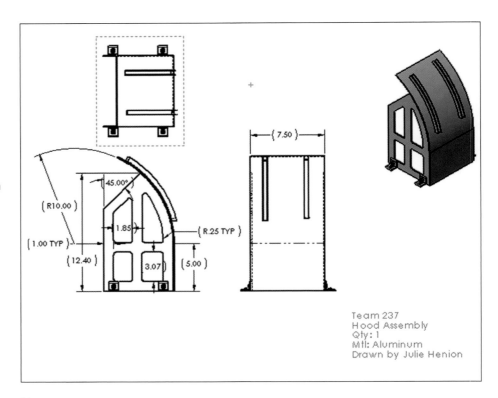

Team 237
Hood Assembly
Qty: 1
Mtl: Aluminum
Drawn by Julie Henion

↑↑ A backing plate and hood create a channel for the scoring balls to pass through. The distance between the wheel and plate is optimized to provide firm contact with the ball. The exit arc of the hood determines the ball's trajectory and is therefore an important design specification.

# ⇄ Optimizing the Shooting Mechanism

The shooting mechanism was created using an iterative process where ideas were proposed, designs prototyped, and improvements made. Refinements were made to each prototype to identify critical performance factors and optimize the overall design. The chosen design consisted of a single wheel that compressed a ball against a flat plate. Optimizing the components of the shooting system required detailed experimentation and testing to make the simple system work effectively.

Experimentation revealed that the ball's exit velocity was a function of the ball's rotational speed as it entered the shooting system. The slowest exit-velocity speeds resulted when the ball had either no spin or spin opposite to that imparted by the shooting wheel. In these instances, energy was lost as the ball rotated along the entrance-to-exit path. The highest exit velocity speeds resulted when the ball entered the shooting system with the same rotation as it exited the shooter, since no energy was lost reversing the ball's spin. This finding dictated the rotation direction for the ball collecting and conveyor systems.

The optimum amount of ball compression was also determined by experiment. The team found that too little compression prevented the ball from making strong contact with the roller wheel and too much compression overloaded the motor. Testing revealed the optimum amount of ball compression to be 1 inch (2.5 cm).

The power required by the shooting system was studied and monitored to maximize performance of the launching mechanism. The team discovered that a single motor could propel a ball but that the system lost a substantial amount of inertia with each shot. The single motor system required a long time to return the system to launch speed. To shoot balls without a delay, two drive motors were used in the final shooter.

Shooting was the highest priority for the team. Through testing, the team realized that if the robot were being pushed by another robot and the driver was pushing back to maintain field position, the shooter motors could not deliver enough power to throw a ball at full speed. To counter this problem, the robot software was modified to limit drive train power during shooting. Since the system was designed to maintain aim while the robot was moving the sacrifice of drive power for shooting power was a rational one.

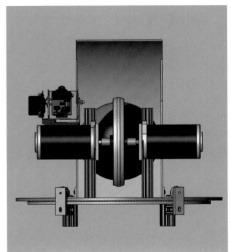

⇄ ↕ Testing verifies the correct spacing of the shooting hood. In addition to the spacing between the shooting wheel and the hood, the side-to-side spacing of the hood is important to keep the ball centered on the shooting wheel.

A belt-driven conveyor system transported balls from the floor to the shooting mechanism. The design used two belts, each driven by a single motor, to push balls against a backing plate and move them along a serpentine track. The motor speed and conveyor-pulley size were selected to move the balls from the floor to the shooter in less than two seconds. Because the foam density of the balls was not consistent, the spacing between the conveyor belt and backing plate was adjusted to compress the balls and not overload the motors.

The conveyor system also served as a storage area for the balls. As each ball entered the conveyor it would advance through the conveyor path and wait if the path was blocked by balls waiting to be fired. The cord that served as the conveyor belt was selected to have enough friction to pick up and advance balls, while allowing for slippage between the belt and balls when the magazine was filled.

⬇⬇ A serpentine path keeps the balls in a line as they progress from the ball collector to the shooter. It's important to keep the foam balls aligned as they can easily jam if given the chance to ride over one another.

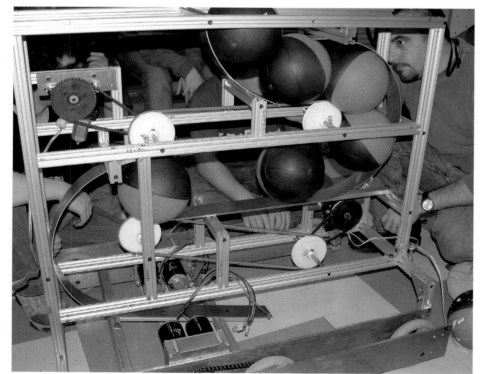

⇄ A supporting frame serves as a mount for the ball guides and as a foundation for mounting motors and pulley shafts. The balls are advanced using a rubber cord that acts as a conveyor belt.

# ⇄ Secrets of Success

In addition to an ingenious design, Team 237 was successful as a result of their careful planning and integration of subsystems thorough testing and incremental improvements to each component. The team's detailed computer models of all components greatly benefited the design effort. The computer models allowed the team to closely examine design details such as fit and function and to avoid problems before they were encountered in the physical world. Individual system models could be integrated, such as with the shooter and the turret, to ensure compatibility between systems. The students' proficiency with the software accelerated the team's ability to transform ideas into reality.

A methodical design approach of sketching, prototype testing, and improvement was applied to each subsystem. In each iteration cycle of the design process, greater

knowledge about system dynamics and response was obtained and the design was improved.

The collective improvements produced a robot that effectively accomplished the team's goal of being able to shoot at the goal no matter what position the robot was in. The turret produced a shooting system with a fast reaction time that allowed the camera to be used as a sensor for closed-loop control. With the goal target located, the shooter locked on that location and the team could fire balls at will into the goal.

Through a process of detailed analysis and careful review, Team 237 converted their initial brainstorming ideas into a highly effective robot that could simultaneously collect and shoot balls while moving around the field and dodging all rivals. Their creativity and imagination combined with careful machining and testing to produce this triumphant robot.

⇊ Sufficient detail in the model is essential for the CAD model to serve as a realistic design tool for planning, analysis and manufacturing.

⇄ ⇊ Once assembled, the intricacies of the robot systems become less apparent but are nonetheless important. The design process starts with an idea, and that idea is continually revised and improved as it progresses from simple sketch to completed product.

# VERSATILITY: A WINNING STRATEGY

## IN PURSUIT OF THE IDEAL ROBOT

When it comes to a winning strategy, the ideal robot had the versatility to master any part of the game, in driver control or autonomous mode, and could be a threat on offense and defense. Such a robot would complement and strengthen any alliance, augmenting the ability of its alliance partners and capitalizing on an opponent's weakness. Having multiple options for playing the game would be an additional advantage for the ideal robot. Turning such desires into a functional machine was the challenge faced by every team in the FIRST Robotics Competition.

Team 294, Beach City Robotics from Redondo Beach, California, is known for developing sound strategy. In 2006, its strategy called for an adaptable robot capable of scoring with multiple methods.

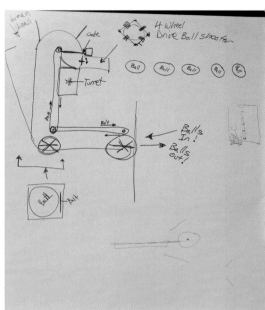

⇆ The concept for Team 294's robot is a result of much brainstorming. The members present their creative ideas to the entire team.

⇊ A preliminary sketch positions the scoring and ball handling systems on the robot. A shooting mechanism is located on the top of the robot.

## ⇉ Meeting The Goal: One Ball at a Time

The upper goals located at the ends of the playing field offered the possibility to quickly accumulate points. But that possibility was only reached when a difficult task was accomplished: repeatedly shooting balls into the circular goal positioned 8 1/2 feet (2.6 m) off the playing field floor.

Team 294's solution to this challenge was a shooting mechanism. The shooter consisted of a 13-inch (33-cm) -diameter solid aluminum disk that compressed the balls against a backing plate. An aluminum hoop was mounted perpendicular to the spinning disk to align the ball on the shooting wheel. The shooter was mounted high on the robot to decrease the chance of other robots blocking the launched balls. The

team was careful to use lightweight materials for the components located high on the robot to maintain a low center of gravity and prevent tipping.

The shooter wheel could consistently fire the balls at the maximum allowable velocity of 39.4 feet (12) meters per second. The shooting wheel speed was measured to maintain the angular velocity needed to propel the balls at the maximum allowable speed. To measure rotational speed, a sensor was mounted near the wheel and a marker was placed on the wheel. Every time the marker passed the sensor a pulse was detected to signify the wheel completed one revolution.

The measurement signal was sent to the robot controller that used the time between signal pulses to determine the shooting wheel speed. When the wheel slowed down, the robot controller detected this condition and a pulse of

power was sent to the wheel drive motor to bring the system back up to speed. This monitoring and control system enabled the shooter to maintain the same launch speed for each ball – an important factor that sustained the accuracy of each shot.

The robot received balls from the human player who filled the hopper at the top of the robot, or from a rotating bar that pulled balls off the floor into the robot. The balls were delivered to the shooter via a conveyor belt. A two-level revolver/agitator organized the balls from the hopper through the use of a spring-loaded trap door; this prevented jamming and ensured that the balls were delivered in a linear fashion. The conveyor also ran in reverse, shooting the balls from where they were collected to score in the lower goals located in the corners of the playing field.

## ⇒ Ready, Aim...

Shooting balls into the upper goal was a significant challenge that required the team to simultaneously aim, maintain field position, and manipulate stored balls in the midst of being defended. Typically, shooting robots were subjected to a constant barrage of bumps from opposing robots playing defense. The best robots countered such harassment with an ability to shoot "on the fly" and score while the robot was in motion.

Team 294 mounted their shooter on a turret that could rotate 180 degrees. This arrangement avoided using the steering system to aim the shooter and allowed the robot to shoot from either side.

The CMUCam2 camera, provided in the Kit of Parts, was mounted on the turret at the highest point on the robot to maintain accurate aim at the goal. The camera and shooting system were rigidly connected, and both systems rotated as a composite unit. Thus, when the camera was tracking the goal the shooter aimed at the goal. The combined tracking and aiming functions allowed the robot to undertake other tasks, such as driving or collecting balls, while scoring in the upper goal.

The camera controls were programmed to correct for the offset in the camera output resulting from the physical separation between the camera and the shooter. The control algorithm automatically pointed the camera at the target goal when the light was located. An optical signal alerted the driver when the target was found, and the tracking system locked onto it. With this feedback from the robot, the driver knew when the shooter could be activated and balls could be launched into the goal.

⇒ A front view of the chassis shows the open entrance where balls are collected from the playing field floor and guided into the robot.

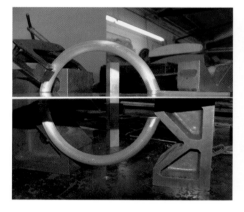

⬇⬇ ⇒ An aluminum hoop guides the balls over the shooter wheel while providing the necessary pinch to impart backspin on each foam ball.

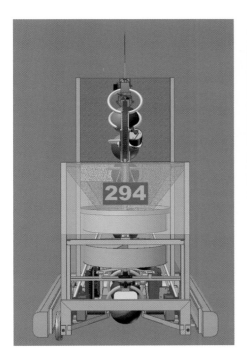

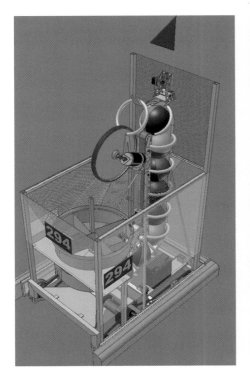

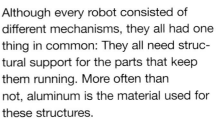

↑↑ ⇄ A CAD drawing of the completed robot and an angled view of the actual machine show the balls collected from the floor being moved into the robot via a conveyor system which can be run in two directions.

↑↑ The motor-driven shooting mechanism is mounted at the top of the robot on a turret that allows it to shoot balls perpendicular to the direction of the drive train.

↑↑ A CMUCam2 camera is mounted on the same turret as the shooting mechanism. Thus, the camera aim always matches the shooter direction.

## ⇉ Holding It All Together

Although every robot consisted of different mechanisms, they all had one thing in common: They all need structural support for the parts that keep them running. More often than not, aluminum is the material used for these structures.

Team 294 constructed an aluminum chassis to serve as the foundation for their wheels, motors, and electronics. For the turret and magazine conveyor, aluminum tubing was used to create a "spine" that consisted of a series of ribs connected to the chassis. The ribs supported the magazine conveyor, which moved the balls up to the shooter, and kept balls from falling out of the magazine as they moved toward the shooter. The ribs also provided a clear view of the conveyor, allowing the driver to see how many balls were captured and ready to fire.

## ⇄ Keeping It All Under Control

With plenty of action during the game and many functions performed by their robot, a simple yet effective system was needed to control the machine. During both autonomous mode and regular play, uncomplicated means for monitoring conditions and controlling action improved the team's performance.

Team 294 designed its operator interface to be controlled by only one person, the driver. The driver controlled all robot functions using only the two joysticks provided in the kit of parts. In addition to offering proportional control of four motors based on the joysticks' locations, the joysticks also include thumb-wheels, finger and thumb switches, and buttons on their base. From this collection of controls, the robot was placed in different modes of play to collect balls and shoot them into the goal.

This layout configuration and control sequence saved time and eliminated driver confusion. The drive system motors were the only ones that operated at variable speeds. Because the speed of the ball delivery mechanisms was preset in the software, all the driver had to do was manipulate buttons to initiate most robot actions.

⇊ An aluminum spine with circular ribs guides the balls up to the shooting mechanism. The inner wall of the spine includes a conveyor belt to transport the balls to the top of the robot.

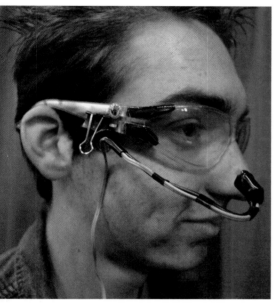

⇈ A driver wears the head's up display during competition to alert him when the camera is locked onto the goal. This saves time by not requiring the driver to look down at the control board and lose sight of the robot.

When in ball-shooting mode, the driver locked the camera onto the goal using a single button: the "lock on" button. This button turned the turret to aim at the goal, and locked on the angle determined by the camera. A heads up display, made of a red light emitting diode (LED) attached to the driver's safety glasses, illuminated when the camera was locked on the goal.

When the LED was lit, the driver knew the shooter could be activated. By depressing the "fire" button, a sequence of actions was initiated: The magazine conveyor belt delivered balls to the shooter, the balls were launched at the goal, and the speed of the shooting wheel was maintained to fire balls at the maximum allowable speed.

A codriver accompanied the driver during the competition to monitor robot functions and exert manual control of individual functions if necessary. By switching to manual control, the pre-programmed sequences and each motor were operated independently. This backup system was needed in case a ball jammed on the conveyor, or if the camera failed and the turret had to be rotated manually towards the goal.

Team 294 used a unique method to create their autonomous mode programs. They simply recorded robot actions under human control and then played back those actions during the autonomous period. The record included instructions for motor speed and timing as previously executed by the human operator. When operating in autonomous mode, the robot played back the recording and mimicked the recorded motions.

Team 294 took the autonomous program a step further. Instead of having a single set of instructions to follow, they pre-recorded 15 different play scenarios. This enabled them to choose the best strategy for the autonomous mode based on their alliance partner and opponent's capabilities, as well as on their starting position on the field. No matter what they faced, Team 294 was ready.

## ⇌ Assembling All the Pieces for an Ideal Robot

Combining effective form and function into one machine can be a difficult challenge. Team 294 identified the needed functions for all aspects of play and built the robot to execute these functions. The robot was dependable and could perform accurately every match. The team's success resulted from a combination of clever mechanisms and careful control.

⬇⬇ ⇌ An AutoCAD drawing and the finished product both depict how form and function are incorporated in Team 294's robot.

The shooter was designed to monitor and adjust its rotational speed if operating parameters were not correct. The turret system accurately located and tracked the goal and was independent of robot chassis' movement. A user-friendly control system was selected and autonomous programs were written for multiple game scenarios. With this collection of strategies, Team 294 turned a complex combination of mechanisms into an efficient machine that consistently performed the actions it was designed to do.

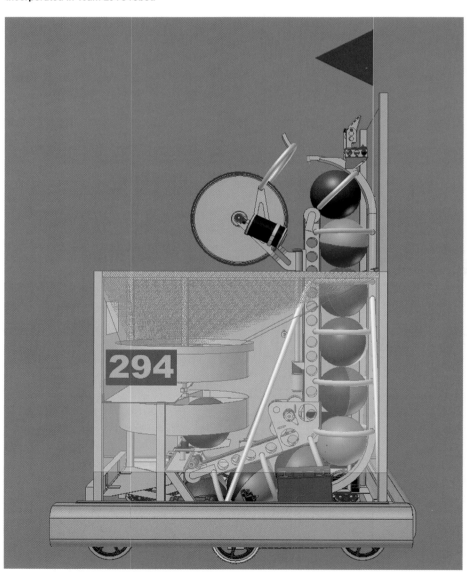

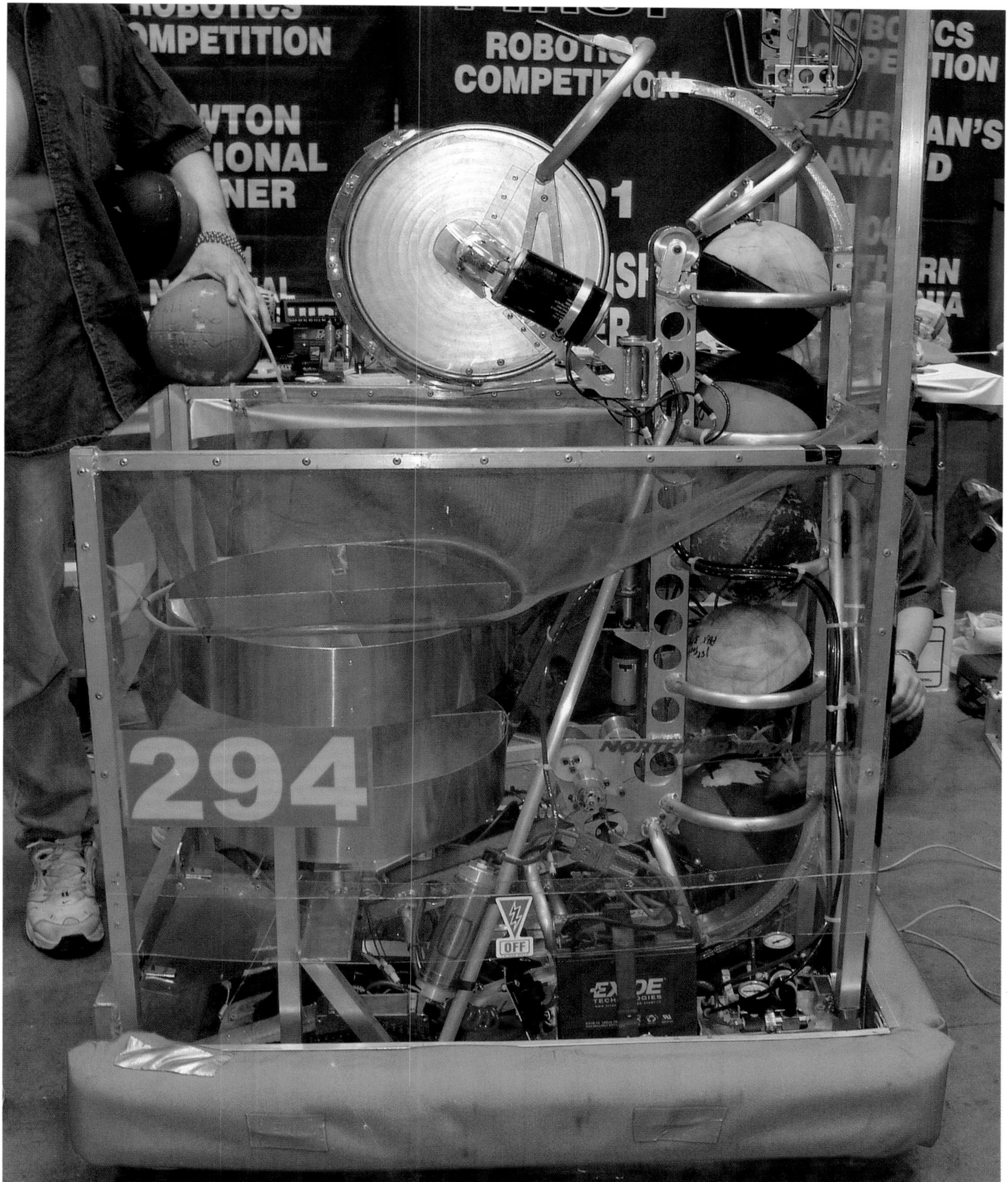
ROBOTICS
COMPETITION
294
OFF

# IGNITING AN INTEREST IN ENGINEERING WITH SPARKY 7.0

## IMPROVING ON PAST DESIGNS

Every year, FIRST teams are challenged to create a robot completely different from any created for previous competition. This variance in games levels the playing field for new and experienced teams as everyone faces a new challenge.

Team 384, from Richmond, Virginia, has seven years of FIRST history and a record of incorporating improvements from each year's competition. The team members name their robots in succession as a reminder of the team's progress each year. Their 2006 robot, Sparky 7.0, is testimony to the continuous learning process practiced by the team.

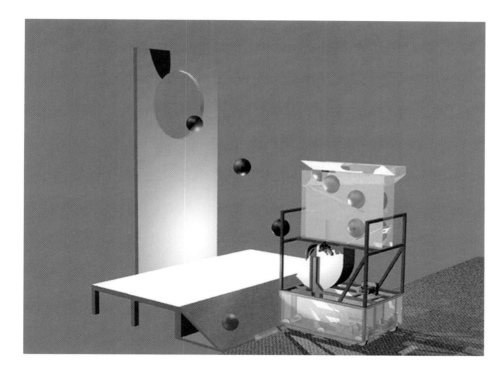

Team 384 designed a robot that was competitive in multiple aspects of the game, from autonomous operation to defensive and offensive strategy. The team's broad range of strengths allowed to pursue a scoring method that complemented its alliance partners. Throughout the competition, the team demonstrated that the best route to success was through teamwork.

## ⇄ Determining the Impact of the Center of Gravity

One quality of Team 384 was its reliance on calculations to estimate the robot's performance. Such calculations, similar to those used in commercial engineering, aided the design process by establishing limits and identifying possible errors before they occurred.

Design parameters can often be established based on estimates. Team 384 estimated the location of the robot's center of gravity (from a calculation) to determine how the robot would respond when mounting the ramp. If the center of gravity was too

⇡⇡ An efficient robot design is the result of learning from past robots. Team 384's robot is a result of team growth.

high, the robot would tip when climbing the ramp.

A graphical method was used to examine the how the center of gravity impacted the robot's ability to climb the ramp. The robot and ramp were drawn to scale, and the location of the center of gravity on the robot's footprint was estimated. A weight vector was then drawn from different heights above the center of gravity.

When on the ramp, the weight vector had to stay within the footprint of the robot to prevent tipping. As soon as the vector pointed out of the footprint, the robot would tip. Using this technique, the team estimated the maximum height of the center of gravity. By keeping the robot's center of gravity below this height, the team could construct a robot stable enough to climb the ramp without toppling over, thus safely scoring the extra points awarded to robots on the ramp at the end of each match.

## ⇄ Improving the Drive Train

Incrementally improving designs was another defining attribute of Team 384. The team relied on its past experience for initial concepts and modified those designs to meet the specific challenges of the 2006 competition, as with the chassis and drive train.

The chassis for Sparky 7.0 was constructed from aluminum, as was Sparky 6.0, but with angles cut out of each corner brace. These angles prevented interference between the chassis' bottom edge and the ramp. The frame was designed to accommodate other robot systems that could be easily bolted on and still fit within the machine's allowable footprint.

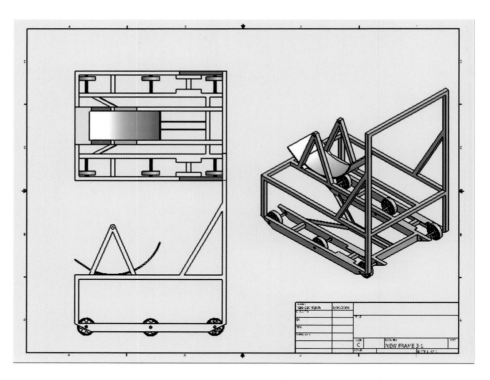

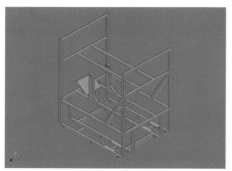

↑↑ ⇄ Detailed drawings help the team build an aluminum frame with angled edges to account for the angle of the ramp. All other robot components are attached to this frame.

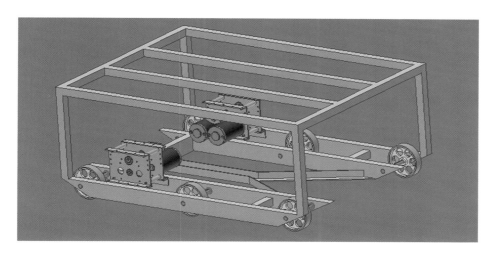

◄ The 2006 robot incorporates a previous drive train design that was slightly adjusted to prevent the foam balls from getting stuck under the robot.

⬇ Two CIM motors connected to a two-speed transmission on each side of the drive train provided Sparky 7.0 with both power and speed to tackle any obstacle on the playing field.

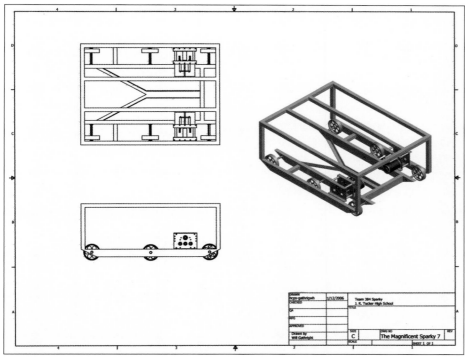

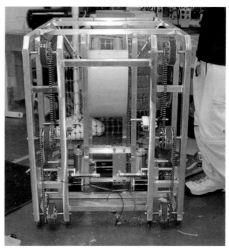

With a limited time to design and build the robot, Team 384 carefully considered its options when determining how the robot would move around the field. Because the drive train for Sparky 6.0 had been very successful, the team chose to adapt that design rather than spend time and money crafting a new drive system. Calculations involving each drive motor's output speed, along with the gear ratio of the transmission, predicted the robot's speed in both high and low gear.

An important improvement was to prevent ground balls from getting wedged in the undercarriage of the robot and prevent it from moving. Small drive wheels with a 2-inch (5.1-cm) diameter were selected to keep the frame near the grand. A small offset was used to lower the center wheel's mounting location. This offset allowed the robot to rotate on just four wheels— a condition that lowered the amount of friction during turns.

The drive train on Sparky 7.0 included a two-speed transmission. The power output from each transmission was coupled with another sprocket to drive all three wheels with a chain. The driver switched the robot into high gear for fast travel across the field. When switched to low gear, the speed was reduced and torque increased. This setting provided enough power to push other robots when playing defense, as well sufficient power to climb the ramp at the end to the match.

## ⇄ Over-Designing and Remembering KISS

The design of the ball hopper was a challenging task for Team 384. To keep the center of gravity low, the hopper had to be constructed with lightweight material. It became a challenge to design a hopper that would be light and still robust enough to withstand repeated impacts on the playing field.

A funnel-shaped hopper fabricated from plastic netting and supported by aluminum braces was first considered, but testing revealed that the balls jammed too easily. Basketball netting was next substituted for the plastic netting material, but further testing showed that this change had not solved the jamming problem.

The next design step taken was the addition of an agitator to organize the balls as they were fed into the launcher. Unfortunately, the weight of the balls applied too much strain on the single agitator, so a second agitator was added to the system.

After taking a step back and re-examining the hopper, the team recognized that it was not following its original design plan of simplicity and robustness. The team went back to the drawing board to design a simpler hopper from scratch. This redesign resulted in a three-level hopper made of lightweight Lexan sheets, connected with rivets, and supported by aluminum angle braces. Each level was sloped, allowing balls to fall from one level to the next by gravity. A feeder regulated the speed of the balls from the ramps to the launcher.

The hopper's three levels had a 10-ball capacity and could be loaded from the top by the human player. A backboard was attached to assist loading. Additional Lexan braces were added to prevent balls from jamming in the corners and when traveling between ramps. Further testing and tuning was conducted to decrease friction and maximize performance.

⬇⬇ A clear Lexan hopper includs internal ramps to linearly dispense the gravity-fed balls down to the shooting mechanism.

ALL 1/4 " LEXAN

| DRAWN bcps-gathrigwh | 1/18/2006 | Team 384 Sparky J. R. Tucker High School | | |
|---|---|---|---|---|
| CHECKED | | TITLE | | |
| QA | | | | |
| MFG | | | | |
| APPROVED | | | | |
| Drawn by Will Gathright | | SIZE C | DWG NO Lexan feed bottom | REV |
| | | SCALE | SHEET 1 OF 1 | |

## ⇉ The Right Combination of Variables

Several ideas for launching balls had been presented while brainstorming. Ideas such as a centrifugal arm and a pneumatic cannon were dismissed in favor of a pitching machine design. Parameters such as the location of the system, the power required to operate the system, launch wheel size, launch angle, and gear ratios were considered. The launching system consisted of two 6-inch (15.2-cm) diameter wheels, powered on the same shaft and running at the same speed. The wheels were placed approximately 2 inches (5.1 cm) apart to allow for two contact points with each ball to increase control and accuracy of each shot.

The wheels were powered by a motor and transmission from the kit of parts and geared to drive the wheel at the high speed needed for launching balls. The launch angle was set by the shape of Lexan feed from the hopper. In retrospect, the team came to regret the placement of the ball launcher. It was situated too low on the robot to accept the gravity fed balls from the hopper. This placement made it easier for opponents to block attempted shots on the goal.

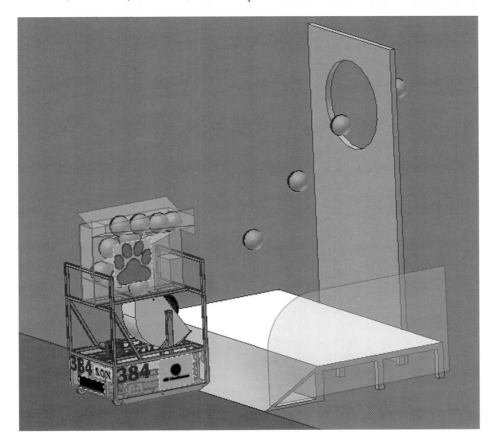

⇄ Although a low shooting mechanism assists in maintaining a low center of gravity, it also means that balls are launched low from the robot, enabling opponents to block the foam balls.

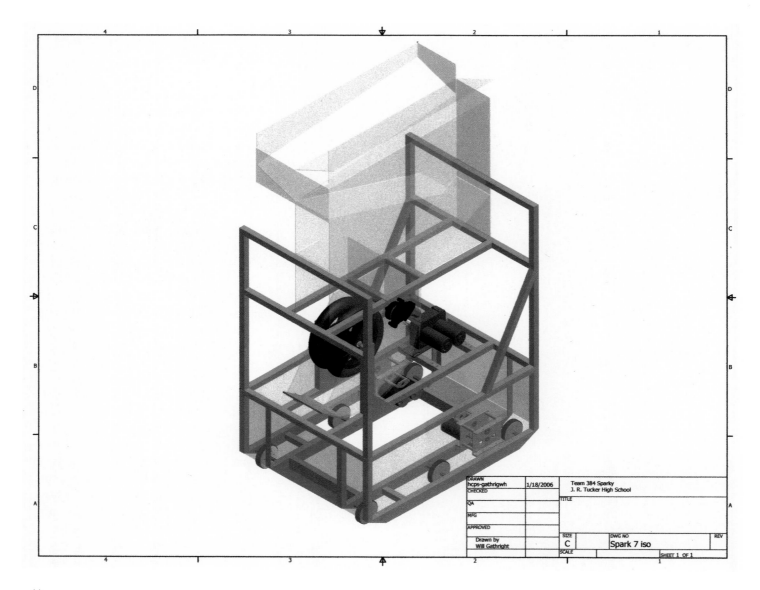

**↑↑** This AutoCAD drawing shows the integration of the shooting mechanism, hopper, ball feed, frame, and drive train.

## ⇄ Component Integration

When the individual mechanisms had been completed, integration became the next challenge. Each component worked fine individually, but to succeed they had to work together. To comply with the dictated weight and size restraints, a mechanism that collected balls off the playing field floor had to be sacrificed.

This compromise ended up benefiting the team as it allowed more room for the remaining components. This example illustrates how simplicity can greatly improve the collective product.

With a finishing touch of bright orange powder-coat paint, Sparky 7.0 was ready to hit the field. By recognizing its strengths, Team 384 was able to best meet the needs of the robot. Time

was budgeted to allow more focus on the mechanisms the team was unfamiliar with, such as the ball launcher; while improvements and upgrades were made to past designs that proved to be successes. This ability to adapt and learn allowed Team 384 to address the challenge of the FIRST Robotics Competition and thrive on the field and as a team.

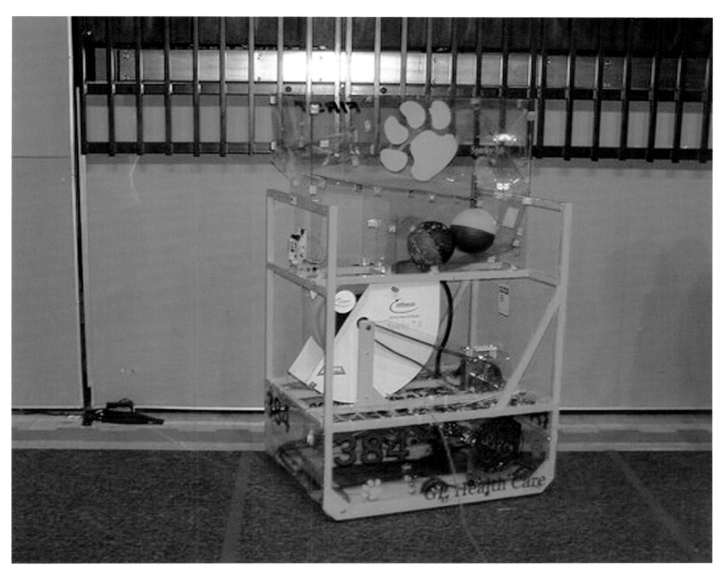

⇑ Sparky 7.0 displayed Team 384 pride with its shiny coat of paint. The final design neatly incorporated all robot components into the required footprint.

# SECTION 03

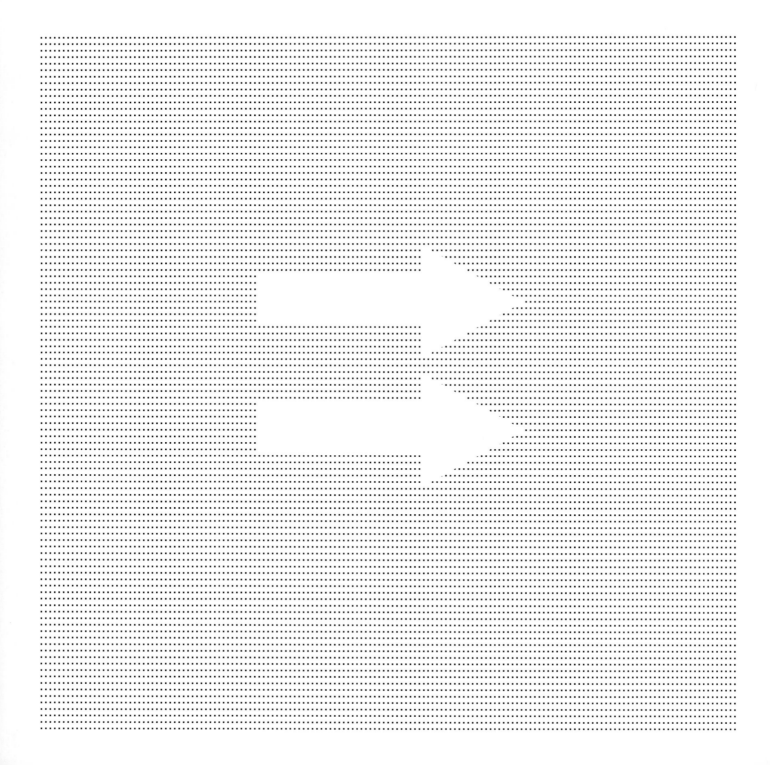

Robot control is an important aspect of performance. The Innovation in Control Award celebrates an innovative control system, or an application of control components, to provide adequate machine functions. Robot functions must be monitored with sensors and measured using the robot control system. Control algorithms use the measurements to determine robot actions and correct for any discrepancies between the commanded signal and the resulting action. Recognition is awarded for performance during the game's autonomous period (where robots operate without human drivers) and during the tele-operated period of the game when operators control robot functions.

In 2006, many teams designed control systems that located the target light above the goal and aligned the robot shooting system with the target. The best of these designs would quickly locate the light and position the robot scoring mechanism with great accuracy. In these cases, the control award recognizes the team's ability to integrate all aspects of the robot design and operation.

# Rockwell Automation Innovation in Control Award

# A SEQUENTIAL APPROACH TO CONTROL SYSTEM DESIGN

## SENSORS, ALGORITHMS AND TESTING

FIRST Robotics Competition Team 111 has a long history of innovative control systems. For the 2006 FIRST season, the team's prowess was demonstrated with a highly accurate automatic aiming system which shot balls into the upper goal. Team Wildstang—a partnership between Motorola, the Wheeling (Illinois) High School Wildcats, and the Rolling Meadows (Illinois) High School Mustangs—used a sequential approach to design their control system. Starting with a concept for both the mechanical and control systems, subcommittees worked independently to design each system, and then they integrated the results.

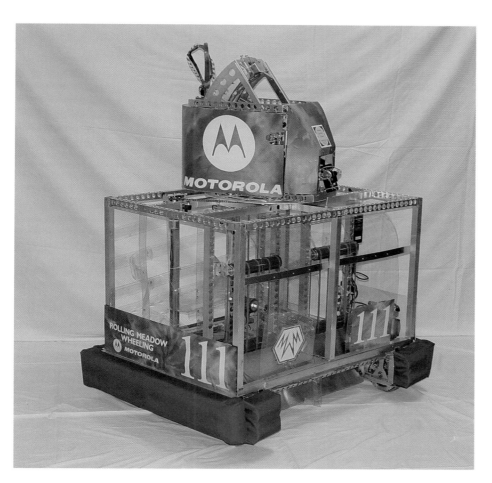

⇅ Though strikingly simple in its external appearance, FIRST Team 111's Wildstang robot is a sophisticated package of cutting-edge technology. Among its many features is a camera-guided turret perched at the top of the robot.

The team first had to decide how the robot should aim the shooter mechanism. While team members had experience with sophisticated drive systems that enabled the robot to move in any direction on the field without changing direction, they decided it would be simpler to have a standard tank-style drive base with a turret-mounted shooter. The team felt that defenders would easily block a fixed shooting system; an automatically aimed turret would be a superior design.

The resulting auto-aiming system, coupled with other control functions, was extremely successful. Once the robot-mounted camera saw the illuminated light above the goal and the driver activated the auto-aim function, the robot automatically controlled the turret orientation and determined the shooter tilt angle to shoot balls into the upper goal. As the robot drove around the field, the turret and shooter constantly adjusted the aim based on data received from the camera. With the shooter locked on the target, the robot driver activated a switch to drive an inner conveyor which supplied balls to the rotating wheel that propelled the balls into the goal.

The control system responsible for this accuracy relied on three components: an array of sensors to monitor robot function and location, a sophisticated control algorithm, and skillful operator control. By working with each robot system independently, the electrical and software control teams developed and perfected their systems before integrating their results with the design and fabrication teams. Coordinating these concurrent activities enabled parallel development of sophisticated processes.

# ⇄ Designing the Auto-Aim System

The iterative process (design, prototype, test, and improve) is commonly used to construct the mechanical components of a robot. Team 111 applied this same sequential design process to the control aspects of their robot, with great results. Their application of this process to design the auto-aiming system illustrated the scheduling and performance advantages that can be gained with this approach.

After settling on a turret-driven shooter, the team divided into three subcommittees (mechanical design/fabrication, electrical wiring, and software control). The mechanical design/fabrication committee designed a turret that was powered by a single motor and a 6:1 gear ratio and which supported the shooting system.

The shooter consisted of a single wheel mounted in a pivoting frame that could be positioned at the required angle to launch balls into the goal from any position on the field. The initial prototype of the shooter was used for testing and later refined to improve performance.

⇊ **Sketching is an important design skill. The similarity between the sketch and fabricated turret base illustrate the value of planning to create and fabricate engineering**

⇊ **An initial layout of components depicts the internal robot frame and the turret base. Drawn to scale, the sketch shows that the frame spacing and turret center are designed to allow balls to pass through the center core of the robot.**

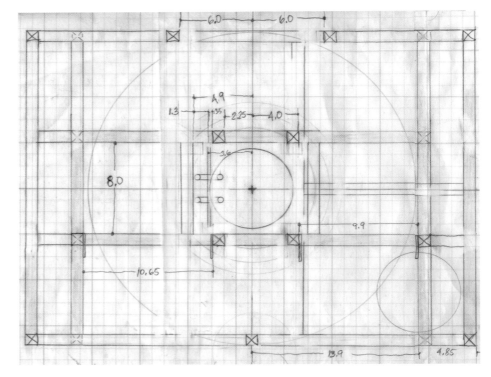

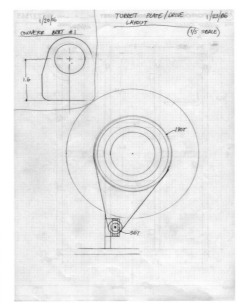

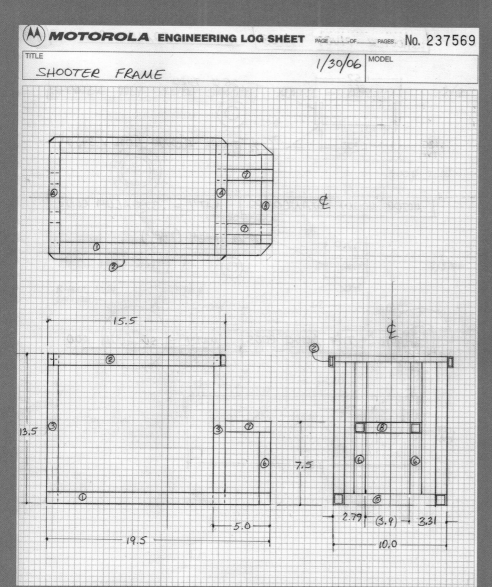

**MOTOROLA** ENGINEERING LOG SHEET PAGE ___ OF ___ PAGES No. 237569

TITLE
SHOOTER FRAME
1/30/06 MODEL

15.5

13.5

7.5

5.0

19.5

2.79 (3.9) 3.31

10.0

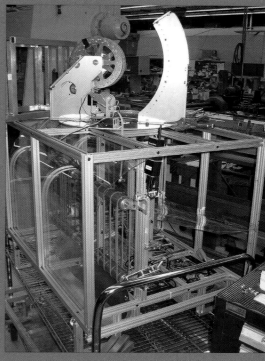

⬆⬆ A shooting-system prototype verifies the plans presented in sketches. Complete integration of the shooting system with other robot components is necessary to evaluate the system and to develop improvements.

⬆⬆ ⇄ An external frame supports the shooting wheel and aiming system. Careful analysis determines the correct dimensions to put enough compression on the bal without overloading the shooting mechanism.

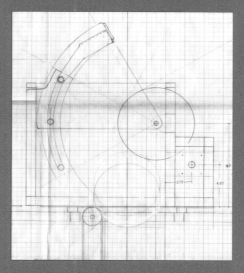

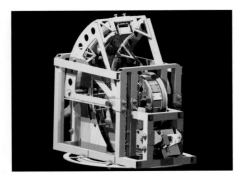

⇆ ⇊ Design details emerge from the computer model of the shooting system and are manifested in the constructed system. The value of careful planning at the earliest stages is illustrated in the dimension marked "3.9" on the bottom plate of the shooter—a dimension that was identified in the early hand sketches of this system.

Based on prior experience, the software control team was confident that it could use the CMUCam2 camera provided in the FIRST Kit of Parts, to locate the goal and control the turret. The team designed a custom circuit with a separate microprocessor as an interface between the camera and the controller. This circuit could independently process sensor signals and send the results to the robot controller, thereby not burdening the robot controller with memory-intensive monitoring and computation functions. In addition, the custom circuit allowed for easier testing and debugging.

The software control team also knew that a variety of other sensors could be utilized to monitor and control important functions. Half of the software team focused on camera control, while the other half worked on code that applied the processed camera information to drive the motors that positioned the turret and shooter.

The first test of the automatically controlled turret aiming system occurred during week three of the build cycle. For this test, the software control team mounted the camera on a prototype turret. The camera output was fed to the custom circuit, which in turn provided input to the robot controller—each of which contained software written by the team for automatic target tracking. At this stage, the team's software could keep the turret aimed at the goal as the robot base was wheeled around the field—an asset that needed refinement and additional testing.

FIRST Robots

A team-designed signal-processing board interprets output from the robot's camera to determine motor control functions. The calculations required to process the camera signal are substantial, and processing the data on the separate microprocessor lightens the load for the robot's main controller.

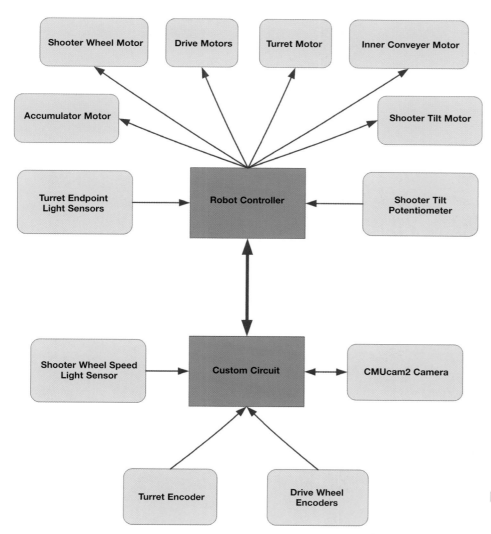

An interface between the custom circuit and robot controller allows each unit to interpret and process data efficiently. The addition of the separate processor prevents the robot controller from becoming overloaded with data processing functions.

The camera is a key component in the Wildstang control system. In its earliest stages, the camera is connected to the turret by the mounting system supplied in the Kit of Parts.

The non-prototype version of the shooter was soon completed and added to the prototype turret. This system enabled the team to shoot balls using automatic aiming and subsequently to identify areas for improvement. By mounting a laser pointer on the turret, it was discovered that the aim frequently drifted by as much as 6 feet (1.8 m) before settling on the target several seconds later. Two causes of this drift were identified: the camera pan and the camera stability.

Originally, the camera had both pan and tilt servo-motors. This required the camera to first rotate and find the target, followed by the turret rotating to match the camera's direction. The dynamics of the camera and turret were different however, resulting in the turret overshooting the target and then oscillating until the camera and turret were aligned. This problem was eliminated by removing the camera's pan servo-motor and using only the turret for camera rotation.

⬇⬇ Assembly of the final version of the shooter with the robot base allows the team to test the system in the configuration that will be used in competition. Every change to the physical system is associated with a change in robot performance, making it necessary to test and retest the final configuration.

The team also discovered that the original camera mount was not sufficiently stabilized, causing the camera to vibrate when the robot and turret were moved. These vibrations were eliminated by designing a new mounting bracket to hold the camera.

Both of these solutions illustrate the integration of software solutions and mechanical design needed for accurate control.

⬇⬇ Since the original camera mount was subject to vibration, a sturdier mount connects the camera to the shooting frame. Rotation of the camera is provided by the turret itself, eliminating the need for a panning servo-motor on the camera.

↑↑ Optical sensors register a signal each time the black mark passes. By measuring the time between these generated signals, the rotational speed of the shooting wheel can be measured and regulated.

## ⇄ Improving Shooting Accuracy

With the refinements to the automatic aiming system completed, the team could turn its attention to two other areas that affect shooting accuracy: ball speed and shooting angle. Through testing, the team realized that if the shooting motor ran at maximum power the wheel speed would vary as the battery voltage decreased during the course of each match. To prevent the resulting variations in speed and trajectory, the rotational speed of the shooting wheel was operated at a constant speed below its maximum and independent of the battery voltage. To accomplish this, the rotational speed of the shooting wheel had to be measured and controlled.

An optical sensor mounted on the shooter frame sensed each time a black mark on the shooter wheel passed by. The sensor was wired to a custom circuit that measured the time between signal pulses from the rotating black mark. This time period was converted to a rotational speed (in revolutions per minute) that was in turn passed on to the robot controller.

↑↑ Sensors are important tools to measure robot performance. A potentiometer mounted on the shooting system measures the tilt angle of the backstop and guarantees the repeatability of shots at the goal.

The controller would adjust the voltage delivered to the shooting wheel motor to consistently maintain 2,500 revolutions per minute.

The last adjustment to the shooter was to program the robot controller to obtain the correct shooting angle based on the distance to the goal. To achieve this angle, the robot was placed at predetermined distances from the goal in 3-foot (0.9 m) increments, and the tilt angle of the shooter was manually adjusted until the aim was correct. The tilt angle, measured by a potentiometer, was recorded in a spreadsheet for each distance.

The camera tilt angle at each distance was also recorded. In competition, the camera tilt angle determined the distance to the goal. This data was then processed to determine the required shooting angle as recorded in the spreadsheet file. A motor on the back of the shooter was activated to place the shooting system at the desired angle.

## ⇉ Human Control Allows Automatic Control

A combination of human and automatic control produced a machine with dead-center accuracy. Testing revealed that the camera had a tendency inadvertently to lock onto the wrong target and begin tracking that object. To reduce this possibility, a decision was made to have the drivers initially aim the robot in the direction of the green target light. When the light was visible to the camera, the drivers engaged the auto-aiming system.

Three indicator lights mounted on a telescoping tube positioned in the driver's peripheral vision signaled to the driver that the target light was located by the camera. These signal lights could be seen without taking one's eyes off the robot and allowed the operators

to drive the robot while monitoring the camera's target-tracking results. A yellow light indicated the camera had detected the target light, a blue light indicated the turret was in auto-tracking mode, and a red light indicated that all systems (shooter aim, rotational speed, and shooter angle) were within acceptable limits to fire.

The custom circuit kept track of the target light location within the camera's field of view. If the camera lost sight of the target—for example, if the robot was pushed off target or a taller robot blocked the camera's view—the custom circuit would rotate the turret to the location where the light was last detected. If the light was last seen on the left side of the camera view, the turret would rotate to the left until the light was once again visible.

⇈ The camera and robot feedback combine to deliver shots into the goal.

⇈ A head's up display of three small lights alerts the driver that the target light is detected, the turret is tracking, and all systems are ready to launch balls.

## ⇄ Control Bonus: Autonomous Mode Success

In addition to the camera and potentiometers, the robot was equipped with other sensors to measure turret angle, turret position, and drive wheel speed. Sensor output was monitored by the custom circuit and robot controller and evaluated to determine motor functions. Control algorithms were developed to provide closed loop control of the turret, shooter angle, and speed.

The effectiveness of the sensor and control system was most apparent during the autonomous operation period.

When operating autonomously, the robot drove to a predetermined position on the field, automatically aimed the turret towards the goal, and shot ten balls into the center goal. Variations in the robot starting location, driving path, and starting time (to allow other robots to move and avoid collisions) were included in the autonomous program. The robot operators set these variations, and the selected program parameters were indicated on a remote display board mounted next to the robot controls.

## ⇄ A Winning Strategy

Integrating the mechanical design and control design was a high priority for Team 111. By doing so, the team was able to focus on control functions from the very start of the design process. This allowed the entire team to accommodate control-driven robot functions and features, such as the location of sensors and the appropriate gear ratios for drive motors. By prototyping control features, testing systems, and making sequential improvements to each robot system, Wildstang produced a well-designed and accurately controlled, high-scoring machine.

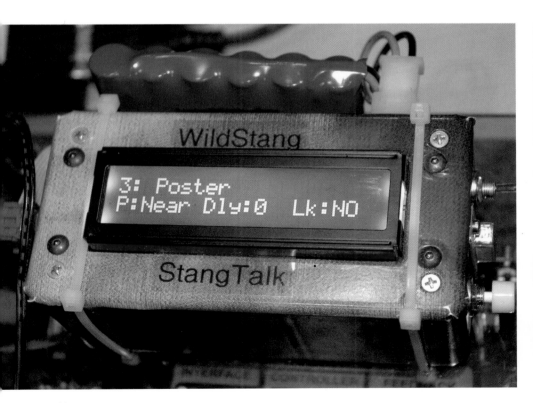

⇈ There is a constant need for information about the robot's readiness. A small display board interprets the settings of dials and switches and informs the operators about the conditions the robot is ready to execute.

⇈ The control system is as important as the robot itself. A high degree of coordination between the robot and the operators is required for effective play and for strong results on the competition field.

# SENSORS CONTROL THE SHOW

## REALIZING HUMANS CAN'T DO IT ALL

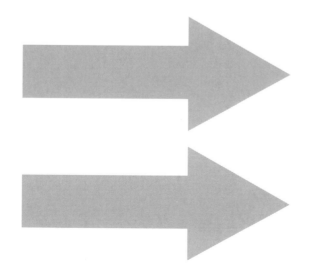

Getting a batch of balls into the lower goals proved a bit harder than one might imagine. Given that the corner goals were 54 feet (16.5 m) away from the human operators, 6 1/2 inches (16.5 cm) above the floor, and at the end of a 24-inch (61 cm) -long, 15-degree slope, the target was relatively small. The five other robots and stray balls on the field further complicated the problem of quickly scoring in the lower goal.

Team 225 realized that given all of the field obstacles, an integrated system of sensors and control was the only reliable method to accurately and consistently score balls in the lower goal. According to philosopher John Dewey, "a problem well-defined is a problem half solved," so in this case Team 225, from York, Pennsylvania, was halfway done with its robot design.

The process began with a brainstorming session to identify machine attributes and functions. The resulting list was reviewed and prioritized to specify desirable robot components such as a four-wheel drive system, a 16-ball capacity hopper, and control system needs. The team decided the robot's primary function was to score points in the lower goal. To support that function, the robot needed an efficient ball pick up, storage, and delivery system.

A conceptual sketch of the machine was created based on the team's list of desirable features. These concepts were refined in drawings produced in the team's CAD lab. The machine's parts were then fabricated from the drawings. Then, the parts were assembled and tested, with a monitoring and control system created to transform the independent systems into a robot.

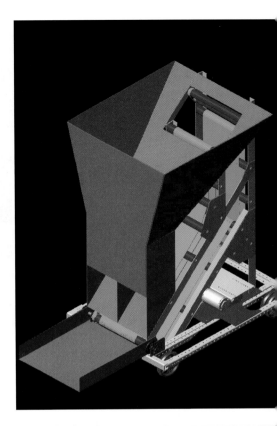

↑↑ The first sketch is an exciting development in the life of a robot. Such a sketch usually follows a discussion of robot criteria and constraints, and the determination of robot functions.

↓↓ ⇒ Progressing from a quick hand sketch to detailed computer models refines and aligns the robot design. The computer models add definition and force-size constraints and system integration concerns to be honored.

↑↑ Once a plan is clear, parts can be manufactured. Machines ranging from hand drills to computer driven mills are used to make robot parts.

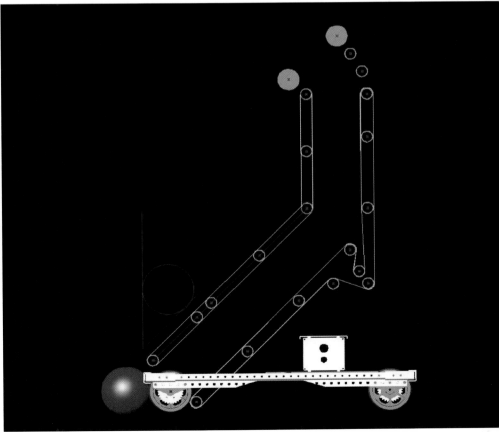

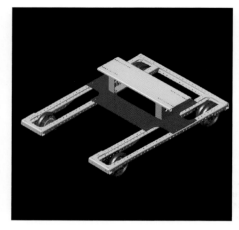

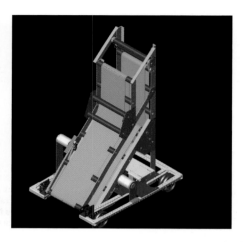

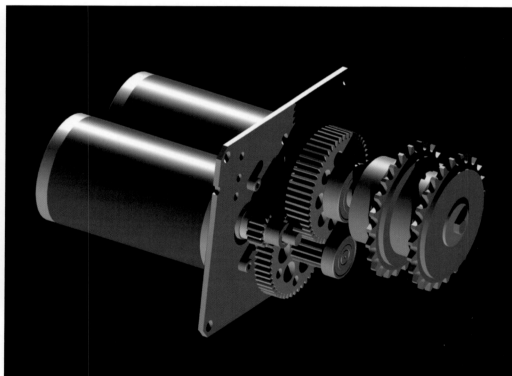

Computer models allow the design to be developed as a series of components that are virtually integrated. Here, the design begins with the robot base and a superstructure is added. Conveyor belts are added to the superstructure to complete the design.

➡ Models of the drive system facilitate integration of the motors and transmission with the wheels. An accurate computer model specifies the chain path between the components.

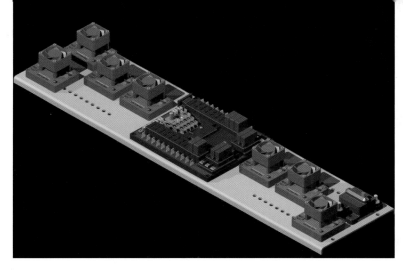

⇇⇊ Even the electrical distribution system is modeled to ensure the arrangement is compatible with the rest of the robot. The layout plan is an installation template and can be used to plan wire paths between the fuse panel and individual motors.

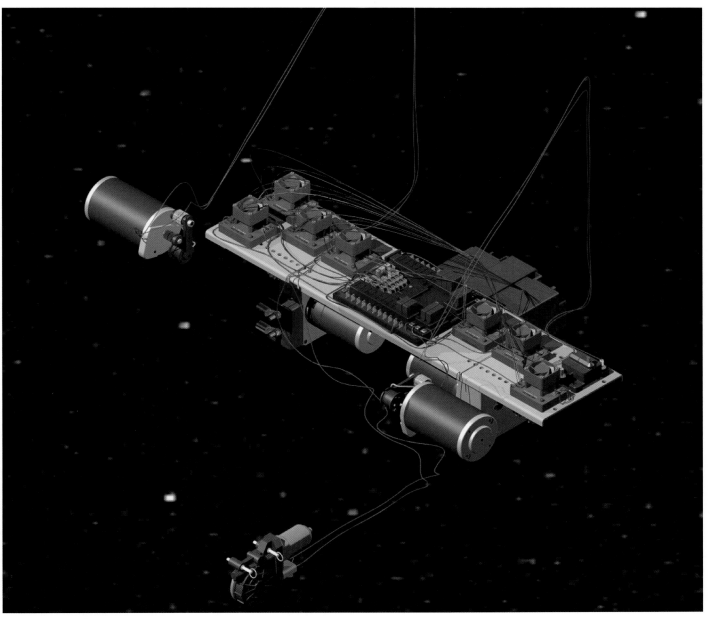

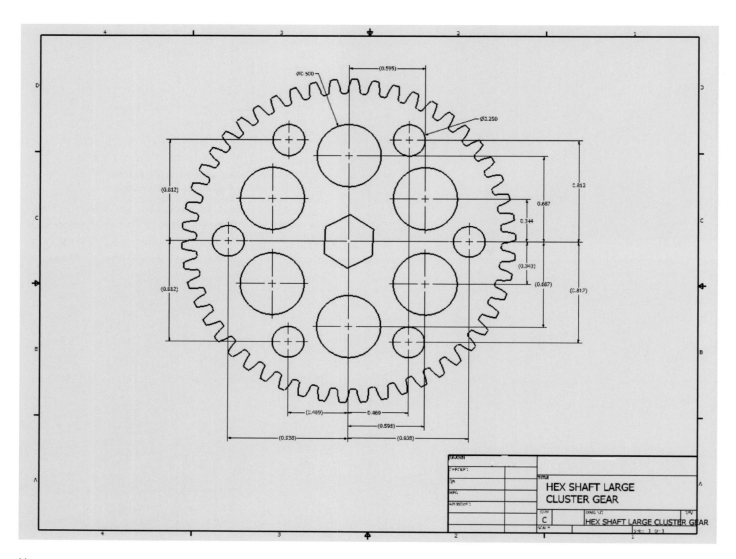

| | | TITLE |
|---|---|---|
| DRAWN | | |
| CHECKED | | HEX SHAFT LARGE |
| QA | | CLUSTER GEAR |
| MFG | | |
| APPROVED | | |
| | C | HEX SHAFT LARGE CLUSTER GEAR |
| | SCALE | SHT 1 OF 1 |

⇈ In addition to serving as a planning tool for layout, computer models also yield construction plans that are necessary to fabricate parts.

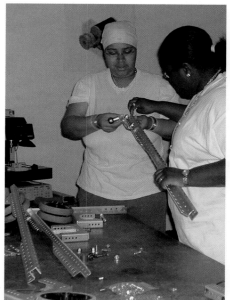

⇉ Team 225's robot base and drive system is constructed from material supplied in the Kit of Parts. The components are designed for FIRST teams and can be assembled using hand tools.

## ⇄ Sensors Make a Machine a Robot

The task of scoring in the lower goal was automated using a collection of sensors and control algorithms. The only task required by the robot operator was to square the robot with the goal and pull a trigger on the driver's joystick. From that point, the control system took over and specified the robot's travel speed, detected the presence of the aluminum ramp in front of the goal, stopped the robot at the top of the ramp, and opened the hopper doors to release the balls. At the top of the goal ramp, the control system prevented any further forward movement of the robot to avoid violating the game rules which restricted robots from entering the goal opening. Five different sensors were used to achieve these functions.

Robot speed was measured with Hall-effect sensors mounted inside the drive train transmissions. These sensors were mounted within 1/4-inch (6 mm) of the main gear in each transmission. The sensor produced a magnetic field that was disrupted by each tooth of the rotating gear. The disruptions in the magnetic field were detected by the sensor and correlated with the distance the robot moved. Since the control system recorded the time between measurements, the robot's velocity was also determined from the Hall-effect sensors. Because slightly different signals were created for each direction of rotation, these sensors also indicated the robot's direction.

⇐ Team 225's robot is a machine designed to score balls in the lower goal. An ample distribution of sensors and the associated code to interpret signals and effect actions transform the machine into a robot.

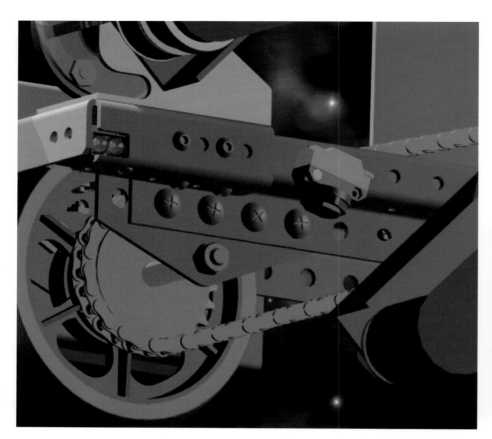

⇄ Optical sensors at the front of the robot shine an infrared light at a 15-degree angle and detect when the light is reflected back. This sensor is positioned to detect the presence of the ramp in front of the lower goals.

⇊ The diamond plate ramp is a strong reflector for the infrared light. When the ramp is detected, an approach program is executed that positions the robot at the goal opening and prevents any further forward movement of the robot, thereby protecting the robot from scoring-zone incursion penalties.

A reflective optical sensor detected the presence of the goal ramp. The optical sensor consisted of two components: a transmitter and a receiver. The transmitter produced an infrared (IR) light beam that could be detected by the receiver. The optical sensor produced a specific voltage level when the IR beam was not detected and a second voltage level when the beam was detected. In this application, the optical sensor was mounted on the robot frame and angled down 15 degrees. When the sensor crossed over the ramp, the IR beam was reflected and the robot control system knew the robot's location relative to the ramp.

With the robot in position on the ramp, a door at the bottom of the collection hopper was opened to deposit balls into the goal. The robot controller measured the electrical current supplied to the hopper door motor with a current sensor that was connected in series with the motor. When the door was fully open, it was prevented from opening further by a physical stop. In these two conditions, the amperage supplied to the motor reached a maximum value, thereby triggering the control system to stop providing power to the motor. By using a current sensor, the team avoided using other sensors, such as limit switches, to monitor the hopper door location. Although limit switches are simple, they are mechanical sensors that wear and can fail. With the current sensor, the team increased the reliability of the monitoring and control system.

A yaw-rate sensor was mounted on the centerline of the robot to measure angular rotation. This sensor produced voltage levels proportional to the robot's rate of rotation. A maximum voltage was produced when the robot rotated in a clockwise direction and a minimum voltage was produced when the robot rotated in a counter-clockwise direction.

By monitoring the output of this sensor, the robot could be commanded to maintain heading and drive in a straight path or turn in either direction at a specified rate.

A two-piece optical sensor was used to detect the presence of balls in the robot. In this sensor, the transmitter and receiver were individual pieces of hardware. By mounting each component on opposite sides of the capture zone at the base of the ball pick-up conveyor, the robot controller detected each ball that entered the conveyor. A control algorithm indexed the conveyor a precise amount to pick up each ball and clear the capture zone. Short bursts of power queued the balls along the conveyor in an orderly fashion that fostered a rapid expulsion from the robot when the time came to score.

## ⇄ Computer Code to Interpret Sensor Data

User-written computer code converted the sensor output data into robot functions. Sensor input was interpreted by the robot controller and used as parameters in the team-written software to direct motor functions.

The signals from the Hall-effect sensors were examined by the robot controller's program to determine the direction of movement for each wheel. A counter kept track of each gear increment to determine the robot's position within 1-inch (2.5 cm) increments. This information was used in the autonomous mode program to direct the robot to specific locations on the field for scoring.

The Hall-effect output was also used to keep the robot in one location on the field and to resist movement from other robots. By measuring the error between the original location and any movement, the robot controller could command drive motor signals to reduce this error to zero and, in essence, keep the robot stationary.

Multiple sensor signals were applied to direct robot functions in a sequential process, such as when scoring. When signaled by the driver, the robot approached the ramp at a slow rate of speed. As soon as the optical sensor detected the ramp, the drive wheel distance was monitored to position the robot at the top of the ramp. Once at the top, the robot would stay in one

⇄ ⇊ Converting sketches into a robot is not an easy task. A design progresses from idea to reality as a result of careful planning, thorough documentation of design decisions, and attention to the smallest details.

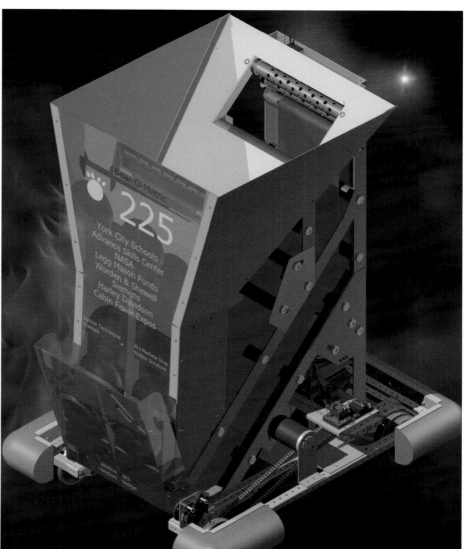

location (by monitoring its wheel rotations to maintain position) and open the hopper doors to release the balls (with the current sensor serving as a tool to stop the hopper drive motor from moving). The conveyor was then energized to deposit any stored balls in the goal and to complete the computer-controlled scoring sequence.

It was essential to test each sensor and the software code before the systems were used in competition. A series of "bench-top tests" examined the system performance. Often, these tests were conducted with the subcomponents in the laboratory before the robot was completed. Since each Hall-effect gear tooth sensor was embedded in the transmission, it was important to make sure the sensors were operating properly before the transmissions were sealed and installed on the robot. An oscilloscope, capable of displaying rapidly changing small-amplitude voltage

↓↓ Balls are stored in the hopper before being released into the lower goal. The vertical fall provides the balls with the momentum needed to roll out of the robot and into the lower goal.

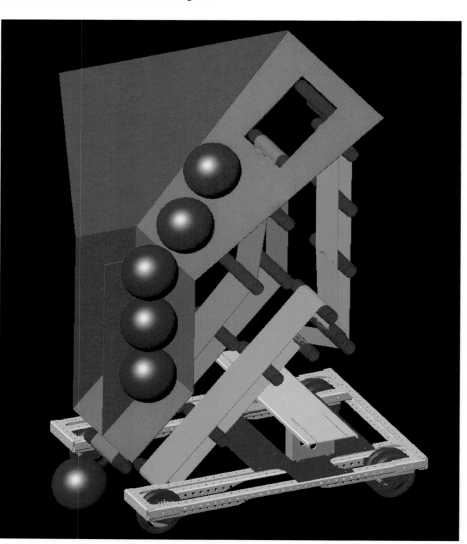

signals, verified the performance of the Hall-effect sensors. When the team was satisfied with the sensors' performance, the transmissions were sealed and installed on the robot.

Additional testing of the drive control algorithm was conducted with the robot on blocks so the wheels could freely rotate without being in contacting the ground. The yaw rate sensor and current sensor were also tested before being installed on the robot to determine the appropriate calibration constants to convert the measured voltages into meaningful values of yaw rate and current.

A second phase of control system testing was conducted when all sensors were installed and the robot was fully functional. A laptop computer was connected to the robot to display the values of all measured parameters and motor conditions. This diagnostic tool allowed the program team to observe the variable values and machine response to ensure the robot behaved according to the coded instructions.

## ⇌ Software Always Behaves as Instructed

Programmers often get to play detective when troubleshooting seemingly errant robot behavior. Such was the case for Team 225 during early operator training when the robot would occasionally stop going forward. To counter the lack of movement, the driver would reverse the machine and normal motion would resume for a time, only to be periodically interrupted by the forward stop glitch.

To determine what was happening, the software team examined their code to find the hidden bug. They discovered that the ramp sensor was always active—a condition that halted movement when the downward-looking optical sensor detected the ramp. This halt was inserted to keep the robot off the ramp when it was not delivering balls which would avoid any penalties being assigned for the robot entering the scoring zone.

Despite being angled 15 degrees to the floor, the optical sensor was triggering each time the robot crossed a taped carpet seam. The tape was reflective enough to signal the sensor to stop all motion. To eliminate this unintended action, additional lines of code were added to activate the stopping function only when the joystick trigger was closed. With this change, the system operated as designed.

## ⇌ The Value of Sensors

This example, combined with the performance differences between sensor-based feedback programs and dead reckoning, time-based coding instructions, illustrate the primary values of sensor-aided instruction: accuracy and repeatability. Because the sensors monitor external environmental conditions such as position, distance, and turn rate, precise maneuvers are assured.

This subtle difference—monitoring and reacting to the environment—is the major difference that separates complicated machines from sophisticated robots. For Team 225, the associated improvements in accuracy and repeatability greatly improved the odds of scoring balls in a goal located 54 feet (16.5 m) away. A task that was hard for a machine to do was repeatedly achieved by this sophisticated robot.

# "SEEING" SOFTWARE AS A CONTROL ADVANTAGE

## VISUAL SOFTWARE TO UNDERSTAND ROBOT FUNCTIONS

National Instruments, a leader in computer-based data acquisition, was one of the sponsors and mentors of FIRST Robotics Team 418, from the Liberal Arts and Science Academy of Austin (Texas) at LBJ High School. With that kind of expertise, it's not surprising that Team 418 was one of the best in the area of control innovations. Not only did the team design a high-performing robot, but also it created a measurement and display system to monitor robot performance, and programmed a variety of autonomous operations to provide versatility at the start of each match. The control and display systems combined effectively to help the team better understand the robot and its operating characteristics.

## ⇌ Camera Aiming for Controlled Shooting

The robot skeleton began as a thin, aluminum angle frame that was riveted and welded to form a strong machine foundation. This material was selected for its easy construction and flexibility in supporting robot subcomponents. In addition, the lightweight structure helped keep the robot's center of gravity low.

Team 418's scoring system relied on a dual set of rollers to trap balls and lift them into a storage hopper. A conveyor belt lifted the stored balls nearly 4 feet (1.2 m), where they were ejected from the robot by a powerful, dual-wheeled shooting device.

The shooter was solidly housed in an aluminum box frame with precision bearings supporting each side of the wheel's drive shaft. A cowling, first prototyped with wood and later fabricated aluminum, served as a back-plate against which the shooter wheels compressed the balls. The complete shooting system was compact, rugged, and protected to prevent damage from opposing teams.

**↑↑** The robot frame is constructed with an aluminum angle, welded and riveted together. This simple frame is the central structure of an integrated ball-gathering and scoring system.

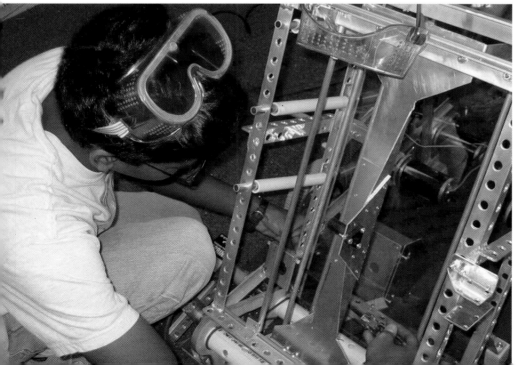

**⇌** The frame, lightened to conserve weight, supports all the motors, rollers, and sensors that transport the balls from the playing surface to the shooting system.

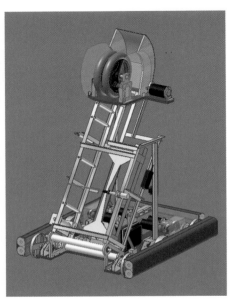

⇄ ⇈ Balls enter the bottom of the robot and ride through the central frame to the shooting wheel. Sensors detect the location of the balls, and monitor robot functions to ensure the accuracy of each shot.

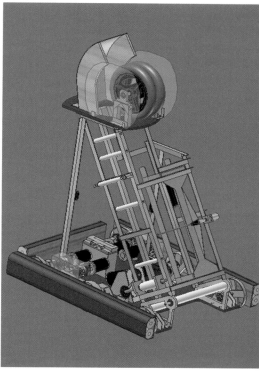

⇄ ⇈ Computer models closely match the final robot configuration. It is much easier to move parts around in a computer model to examine options and ensure fit, thereby eliminating problems during the construction phase.

Aiming was accomplished by rotating the robot with the drive system. A powerful, dual-speed, four-wheel drive system was designed to quickly cover the field and easily align the robot with the goal. The drive and shooting systems were used in tandem to position the mechanisms and propel the balls into the upper goal.

The camera provided in the FIRST Kit of Parts was an important sensor on Team 418's robot. The camera constantly searched for the illuminated target on top of the goal. Once the light was detected, the camera's servo motors kept the target centered in the camera's view. Data on the orientation of the servo motors, when locked on the target, was used to develop a targeting solution for the robot. The camera pan determined the alignment of the robot, and the camera tilt established the distance to the goal.

The distance the shooter could propel a ball was a function of shooter speed—a parameter that could be adjusted to accommodate the robot's distance from the goal. The speed and resulting trajectory were determined from testing. These results were stored in the robot controller's computer memory, and could be accessed to determine the required speed based on the camera's tilt angle. These automated processes—to align the robot and determine the shooting speed—were just two of Team 418's many control innovations.

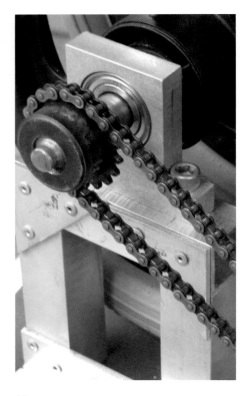

⇈ A stable shooing platform is needed to reduce vibrations and deliver accurate shots into the goal. Aluminum box tube is carefully fitted to form the frame and support the shooting wheel.

⇉ Wood is a useful material to prototype robot systems. The shooter system prototype is constructed with wood to evaluate the optimal spacing between the cowling and the shooting wheel.

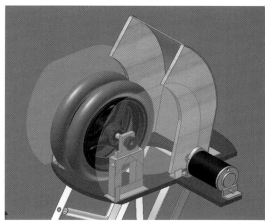

⇄ ↓↓ The wooden shooter base is retained for its strength, but a sheet metal cowling replaces the prototype hood. Located at the highest point on the robot, the reduced weight of the sheet metal greatly improves the robot's stability.

⇄ ↑↑ Each drive wheel is an independent system, from the drive sprocket to the motor and all components in between. This design introduces redundancy and enhances control. Small servo motors, mounted on the motor transmissions, switch the drive system from high to low gear.

# ⇄ An Integrated Control and Monitoring System

Other innovations included the control box and data acquisition system that enabled the operators to control and monitor robot systems. Team 418 determined that while joysticks were useful for driving, the embedded buttons and triggers on a joystick were not optimally arranged for controlling discrete robot functions. To better position these switches, they designed and constructed a control box to make robot operations more intuitive and user-friendly.

A large number of independent controls and autonomous functions were required to manipulate the robot's subassemblies. The control box was designed to organize these functions and provide clear divisions between groups of controls. Controls that might be used simultaneously were placed on opposite sides of the control box to allow for independent operation with each hand. For example, the left hand controlled the upward feed of balls on the conveyor belt while the right hand controlled the manual speed of the shooting wheel. Functions requiring greater precision, such as the shooting speed, were controlled with the operator's dominant hand.

The control box housed a laptop computer that displayed robot data in a graphical format that was easily read. The screen of the laptop was positioned just below the operator's view of the field to allow the operator to scan easily feedback information while driving.

LabVIEW, a powerful graphical development environment for signal acquisition, measurement analysis, and data presentation, was used to process, interpret, and display robot functions. The team used this software to monitor ball position in the conveyor lift, as well as shooting wheel speed, camera servo

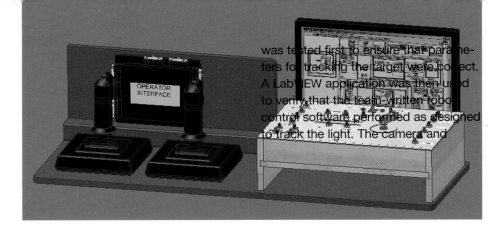

was tested first to ensure that parameters for tracking the target were correct. A LabVIEW application was then used to verify that the team-written robot control software performed as designed to track the light. The camera and

↑↑ In addition to the Innovation First operator interface, the control board includes three other components: joysticks to drive the robot, a button box to execute robot actions, and a laptop computer to monitor performance. A monitoring program running on National Instruments' LabVIEW software oversees robot functions and alerts the drivers to abnormal conditions.

positions, and all other motor speeds. LabVIEW applications were also written to inform the operator of ball-storage capacity, low battery power, and alignment of the robot with the upper goal.

In addition, LabVIEW applications were created to bench-test most of the control software as it was being written, to ensure that a working program was ready when the robot was assembled. This simultaneous development of the robot and its controls enhanced the team's ability to meet a tight production schedule.

As soon as subcomponents were constructed, their control systems were tested. The camera control subsystem

software were both then tested on a prototype drive system to examine the performance of the complete tracking system. Since it had been fully tested, the control system needed only minor tweaks to optimize performance on the final robot.

LabVIEW applications were used to understand the output of each digital sensor, and to test the program logic that converted the measured signals into robot actions. For these tests, sensors were connected to the robot controller and manipulated to simulate robot actions. The robot controller's signals were examined to make sure they produced the intended robot functions. This process facilitated code development and verified the control logic before it was implemented on the robot.

## ⇉ Autonomous Mode Multiplicity

Team 418's robot could also execute multiple autonomous programs. Three separate operations for offense and defense were programmed and ready for selection based on the team's needs for any match. The operations were selected with single-pole, single-throw switches located on the robot.

The autonomous programs included instructions to score points in either the upper or lower goals. The upper goal program was the most complex. At the start of the autonomous period, the camera would search for and acquire the target light. The camera's output was fed to the robot controller, which determined the power needed by each drive motor to align the robot with the goal. The camera's output was also used to determine the necessary shooting speed to reach the goal. Once these tasks were completed, the balls were fed to the shooter and propelled into the goal.

For lower-goal scoring, the robot would be positioned in line with the goal. At the beginning of the autonomous period, the robot would move toward the goal using a timer to gauge the distance traveled. When the robot reached its destination, the ball-feed rollers were energized and the balls expelled from the robot. The speed was carefully controlled to roll the balls gently into the lower goal and avoid any bouncing that would prevent them from scoring.

An autonomous program was written for defensive play. In this program, the robot moved at a high speed to block or bully opponents.

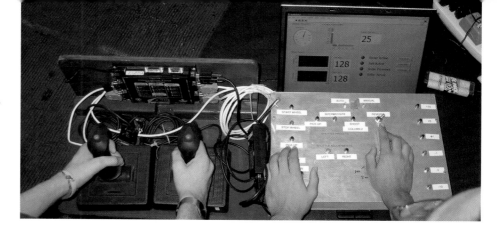

A gyroscope measured the robot's direction, and this feedback was used to guide the robot in a straight line toward its goal to block the opponent from scoring.

## ⇉ Balancing Control

Team 418's control system effectively integrated manual and automatic control. While some robot functions, such as firing balls, were under manual control, other systems, such as the shooting wheel speed, were automated. Graphical displays were used to monitor robot functions and alert operators about significant performance conditions. During hectic matches, important performance information could be obtained by quickly glancing at the display.

Team 418 carefully blended design features and control options. The team's LabVIEW application for system testing proved an efficient tool to speed development. The intuitive and organized layout of the control panel aligned with the operator's preferences and eased human control. The multiple autonomous modes of play increased the robot's utility and provided an ability to select the most beneficial actions for each match. Altogether, the control interface between man and machine fostered better understanding of robot actions and maximized the team's overall effectiveness.

⇈ The driver concentrates on maneuvering the robot around the field while the copilot handles all other robot functions, including monitoring system performance using the LabVIEW display panel.

# AUTOMATIC TARGETING SYSTEM SCORES A BULLS-EYE

## DESIGNING A ROBOT TO OPTIMIZE CONTROL FEATURES

With the unveiling of the 2006 FIRST Robotics Competition, Team 494 from Goodrich High School, in Goodrich, Michigan, understood the significance that robot control would factor into the game. The team anticipated a crowded playing field in front of the goal, with perhaps a concentrated sweet spot near the goal where most of the offensive play would be centered. Speculating a bit, Team 494 viewed the game as one where time would be short, the field would be crowded, and windows of opportunity for scoring would be relatively small.

Given these concerns, the team decided that the fundamental feature of their robot would be an ability to automatically find and track the goal. This feature developed into what was called the Automatic Targeting System (ATS) to enable fast and accurate shooting under the most adverse conditions. The objective of the ATS system was to quickly locate the goal and track it while the robot moved (or was moved) across the field. With this capability, the team envisioned the robot always being able to score points, even when subjected to unintentional movement from opposing robots. The ATS was adopted at the start of the design process as a system requirement, and the robot was designed around this attribute.

⬆⬆ The ATS controls all robot functions to deliver balls into the upper goal. The robot design is based on this control system, with all design decisions made to augment the robot's ability to score.

# ⇄ Building a Robot Based on Control Requirements

The design team began the creative process understanding that ball-shooting accuracy would be the most important robot feature. All design decisions would support the automatic target tracking objective, and it was essential that mechanical systems would not compromise the control system's agility.

The team examined two alternatives for aiming the shooting system: driving the robot and using a turret. They decided the best response would be obtained with a turret-mounted shooter, since that system would be faster and have more versatility when the robot's motion was restricted. Anticipating a crowd near the goal, the team designed a shooting system with a variable azimuth angle to adjust the shooting angle. A quick ball gathering and storage system provided the shooting system with a nearly continuous supply of balls.

With these mechanical systems in mind, the team produced sketches for a robot that could support the ATS. The design called for a robot that gathered balls and used a conveyor belt to lift them to the top of the robot. At that point, the balls would advance to the shooter or be diverted into a helical storage system. The helix design offered an ideal combination of high-density packing and single-file storage. This structure provided an open-center volume through which balls could pass and be forwarded to the shooting mechanism. Human players could also load balls by shooting them into an open net at the rear of the robot, with the balls then entering the helical storage system.

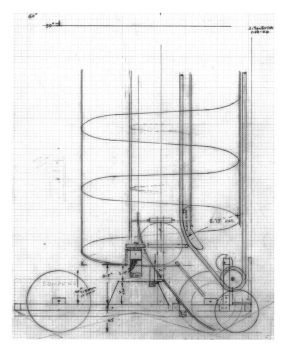

⇄ Accurate sketches model the motion of balls through the robot. Balls enter a central conveyor at the base of the robot.

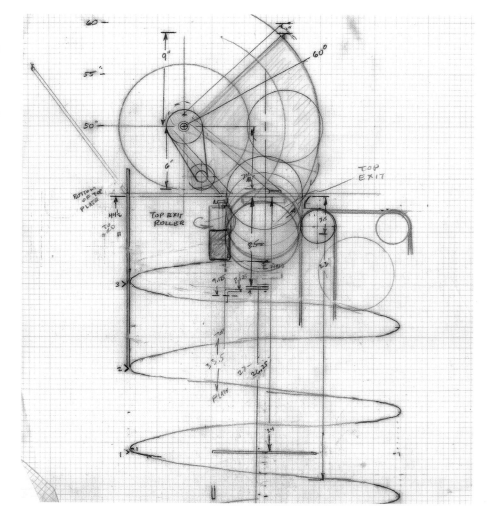

⇒ The balls progress through this central conveyor to the shooting system or are diverted to a helical storage cylinder, where they are kept until needed.

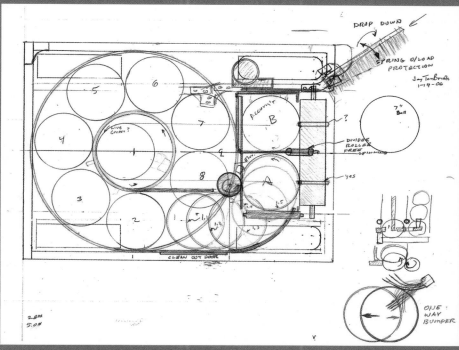

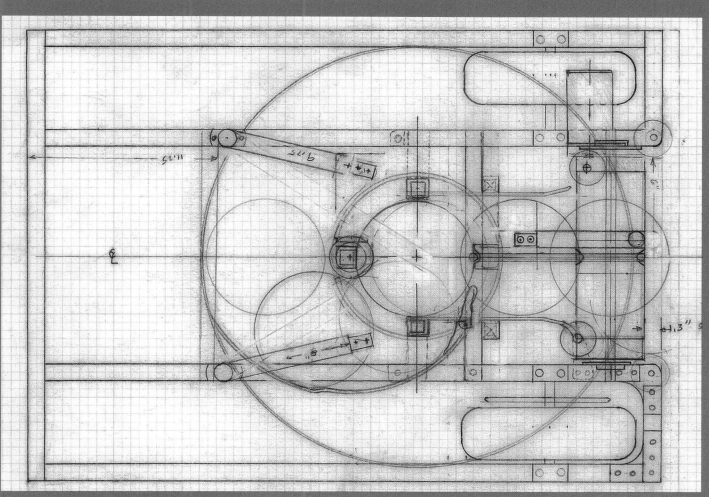

⇆ ⇊ The ball gathering, conveyor, storage, and shooting systems are tightly integrated to fit within the specified design space.

## ⇄ Design Improvements For Control Excellence

To support the ATS, the robot had to be highly maneuverable, the shooting system unrestricted, and the rate of fire as fast as possible. These criteria resulted in a four-wheel-drive, redundant-propulsion system, a fast-spinning turret capable of variable elevation angles, and a lightweight shooting wheel. Each mechanical system was examined and improved to support the ATS, and novel control applications were created to enhance the performance of each subsystem.

The overarching requirement for the control system was reliability. To this end, redundancy was created in the power and transmission systems so that failures in any mode would be mitigated by the presence of a backup system. Three motors were used for power on each side of the robot, with the drive system designed so that if any one motor failed the other two could continue to operate. Individual chains connected the motors to the drive shaft of each wheel to increase the system's redundancy.

The team created an electrical load management control routine that was programmed to continually measure the current draw on each motor. Electrical power delivered to each motor was regulated to minimize the current drawn by each motor at any speed. This form of monitoring prevented the motors from drawing down all of the battery's energy.

During testing, the team discovered that pneumatic drive wheels provided too much resistance during turns, causing the robot to bounce—a condition that would conflict with tracking the goal. To reduce side friction during turns, the pneumatic wheels were replaced with traction wheels and omni-wheels. Further testing found that when both omni-wheels were located on one

**↑↑ Mechanical components controlled by the Automatic Tracking System include the robot drive, ball retrieval, ball transport, ball storage, and launching systems. The shooting wheel is mounted on a turret that is positioned by the Automatic Tracking System.**

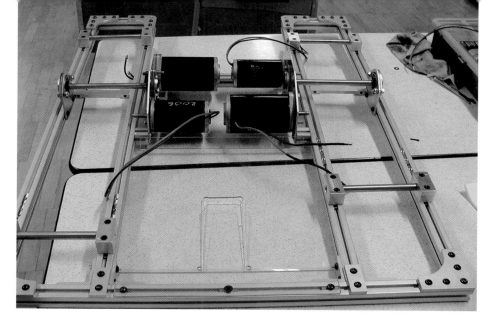

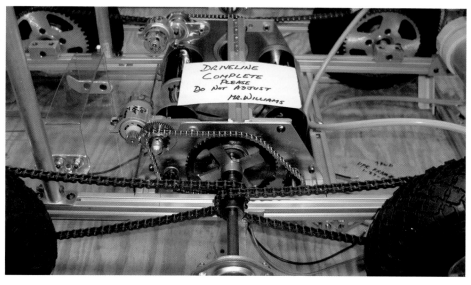

⇈ Supported on a frame constructed from structural aluminum, the robot is powered by a drive system that has a high degree of redundancy. Three motors power each side of the drive train, and the system can operate if any motor fails.

end of the robot, the robot was highly maneuverable but could be easily displaced by another robot pushing it from the side. By locating the omni-wheels in opposite corners, the team was able to retain the robot's maneuverability and resist its being displaced by pushes.

An additional control feature was added to the propulsion system to enable the robot to hold its position during shooting. When needed, the drivers initiated a dynamic braking control system where encoders measured the position of each wheel. Any deviations to the wheel's position were countered by directing power to restore each

wheel to its original position. In this mode, the robot maintained its field position even when disturbed by opposing robots.

The turret was designed to quickly rotate to any position within a 270-degree range of motion. To accomplish this, an output gear on the turret drive motor was mated with the outer edge of the turret. Since a gear pattern was machined into the turret's outer edge, the high-speed output of the motor resulted in a fast but manageable speed for the turret.

A desire to be able to fire balls from any position on the field required controlling the elevation angle of the shooting mechanism. To adjust the elevation angle, a wedge-shaped gear section was attached to an aluminum frame that kept the balls in contact with the shooting wheel.

As this frame rotated back, the balls would be released sooner and have a greater trajectory angle. A potentiometer mounted on the frame's axis of rotation measured the elevation angle of the shooting system. Testing revealed the relationship between angle and

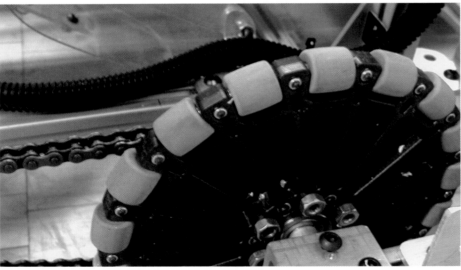

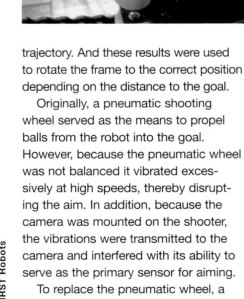

trajectory. And these results were used to rotate the frame to the correct position depending on the distance to the goal.

Originally, a pneumatic shooting wheel served as the means to propel balls from the robot into the goal. However, because the pneumatic wheel was not balanced it vibrated excessively at high speeds, thereby disrupting the aim. In addition, because the camera was mounted on the shooter, the vibrations were transmitted to the camera and interfered with its ability to serve as the primary sensor for aiming.

To replace the pneumatic wheel, a large-diameter aluminum wheel was

manufactured and installed as the shooting wheel. The aluminum wheel was uniform in shape and perfectly balanced, and thus its operation did not disturb the aiming process. A potentiometer was mounted on the shooting wheel to measure the rotational speed, and the output of this sensor was used to regulate the wheel speed.

To conserve the robot's battery during competition, the wheel was rotated at a lower speed than was required for shooting; if the robot needed to shoot the ball, the wheel could be brought up to speed much quicker than if it weren't rotating at all.

The wheel shooter potentiometer was also used to monitor the system's readiness to fire a ball. As each ball was shot, the speed of the shooting wheel momentarily dropped as momentum from the wheel was transmitted to the ball. The control system recognized this speed drop and boosted the electrical power to the shooting system motor. The sudden power boost returned the wheel to the required shooting speed, at which time another ball was fed into the shooting system.

↑↑ The turret drive motor is geared to the outer edge of the turret base and accurate positions of the turret. A potentiometer geared to the turret drive motor measures the turret's angular position.

↑↑ The turret has a 270-degree range of motion to allow the team to score from nearly any location on the playing field.

⇉ Manufactured from a single piece of aluminum plate, the shooting wheel rotates at a high rate of speed but produces minimal vibration. This solid plate replaced a pneumatic wheel that produced significant amounts of vibration when rotating at the high rate of speed required to launch balls.

## ⇄ Team 494's Automatic Targeting System

Team 494's ATS enabled fast and accurate shooting under the most adverse conditions. This comprehensive control system located the upper goal, locked the shooting system's azimuth and elevation angle onto the target light, shifted the drive system into lock-down mode to maintain field position, and brought the shooting wheel up to the speed needed to propel balls into the goal. Any deviations from optimal settings were detected and corrected by the control and monitoring system. A high degree of control was obtained using only two types of sensors: potentiometers and a turret-mounted camera.

The multi-turn potentiometers produced a voltage that varied with the sensor's angle of rotation. Standardized potentiometers were used throughout the robot to measure the shooting elevation and turret angle, as well as the speed of the shooting wheel and each of the four drive-system wheels. Using a common sensor for each measured parameter standardized both the hardware and software. Small details, such as having the electrical leads on each potentiometer be the same length and pre-assembled allowed for quick repairs if damage occurred. With a common sensor, the same software algorithms were used to convert the voltage measurements into angles or speeds, thereby making the final control program more robust and less prone to errors.

The camera served as the principal sensor for aiming the shooting system. Once the operator brought the target light into the camera's field of view, the ATS took control of robot functions. A sequential process converted the original camera signal into a scoring opportunity.

With the target in view, the camera software yielded two key pieces of information: the horizontal and vertical locations of the target light. With this information, the turret immediately rotated to center the horizontal location in the camera's field of view, thereby aligning the shooter with the goal. Any deviations from the desired position were measured by the camera and corrected by supplying additional power to the turret motor.

The vertical location of the target in the camera's field of view assisted in measuring the distance to the goal. This information controlled the speed of the shooter and the shooter elevation angle. Testing determined the necessary speeds and angles to shoot the ball from any location on the field. These results were stored in the robot's microprocessor memory. In competition, when the vertical location was measured, the test data was consulted and the shooting wheel and elevation motors were supplied with the power needed to accurately fire the projectiles. As with the turret, any deviations from the desired positions were recognized and corrected by the ATS.

In addition to the aiming functions, the ATS also engaged the dynamic braking feature that secured the robot's position on the field. Any hostile attempts to move the robot were automatically countered by opposing forces generated by the drive system to keep the robot stationary. By remaining in position, the robot's aim was not disturbed and the robot could accurately shoot all of the balls held in its storage helix.

⬇⬇ **The camera is mounted near the shooting wheel to accurately aim the turret at the upper goal. The camera rotates with the turret, and its elevation angle changes with the turret aiming system.**

## ⇄ Designing a Robot Based on Control Functions

Team 494's machine was designed from the start with the Automatic Targeting System integral to the robot. Its control functions necessitated quick system response and efficient use of sensor measurements. Redundancy was designed into critical components, and the team's use of standard hardware and software reduced the amount of time it took to implement functions and improve operations.

The team estimated that the Automatic Targeting System more than doubled the machine's scoring capability. Rather than needing drivers to precisely align the robot with the goal, the robot was quicker to find the target and accurately adjust its conditions to score. By controlling the robot the team could control the score, much to the appreciation of their alliance partners.

⇄ The robot and its control system are an integrated design. Data is collected and interpreted by the robot controller, and team-written program instructions direct robot functions for fast and effective scoring.

# TEAM 1114

# STILL ON TARGET

## FOUR-WHEELED SHOOTING

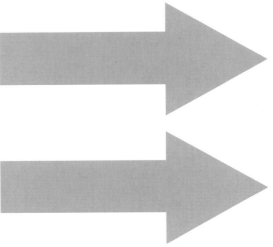

Team 1114 began the design process with a simple question: Do we move or stand still? The question was not about the robot's overall action, but about its function during shooting. This single question guided the robot design and helped the team, from General Motors-St. Catherine's Powertrain & Governor Simcoe Secondary School in Ontario, Canada, think about possible alternatives.

The team decided to use a fixed launcher on the robot to keep the design and controls simple. A fixed launcher required that the entire robot base be used as the aiming mechanism, rather than using a turret to aim at the goal. Recognizing that the robot would need to stay in one position to shoot multiple balls, the team designed a drive train and control sequence to ensure this condition. High traction, tank-style treads were selected for propulsion, with the expectation that the high levels of friction from the treads would also resist motion from opposing robots.

A control algorithm was written to keep the robot stationary while shooting. Sensors monitored the original location of the robot before shooting commenced, and a feedback system kept the robot in position. Referred to by the team as "drive base lock," this condition freed the robot operators from worrying about maintaining position in the often congested area in front of the goal.

By analyzing the trajectory with ball projectile formulas, including consideration of air resistance, the team calculated the necessary tilt angle of the shooter. With a 29-degree shooting angle, the robot could hit every shot at the upper goal from distances between 7 and 27 feet (2.1 to 8.2 m). As such, there was no need to further complicate the design with an adjustable tilt mechanism. By keeping the shooter angle physically fixed and the robot base dynamically fixed, the team created a simple design that was further augmented with effective control strategies.

↑↑ Early in the design process, Team 1114 investigated a four-wheel shooting system. The system performed well.

↓↓ The robot frame is constructed from aluminum plate and bar stock, cut to length and welded together. The design includes foundations and supports to mount motors and other robot systems.

## ⇄ Robot Creation

Team 1114 experimented with a four-wheeled, wood-framed shooting system at the beginning of the robot design process. The design began with an aluminum base, outfitted with four drive motors and multiple speed transmissions, to provide a range of power for different driving conditions. The base was augmented with tread propulsion and a multiple-band ball pick-up system. The shooting system, towering over the base, was positioned at the height needed to shoot over opposing robots. A ball-storage hopper, constructed with polycarbonate panels on a PVC frame, completed the design.

Each aspect of the robot was modeled using computer-aided design software. Even the operator control panel was initially planned using computer-based design tools before it was constructed with real dials and switches. The computer-aided drawings served as a record of the design that was evaluated to examine functions and make improvements.

## ⇄ Control Features For On-Target Shots

Intuitive control was a guiding principle for Team 1114. This strategy, adopted to simplify driving, led to the creation of four control features: drive base lock, autonomous target check, camera aiming, and curved shooting.

Drive base lock was the control algorithm developed to maintain the robot's position and orientation at a specific location on the field. A simple click of the joystick trigger relinquished driver control to the onboard microprocessor. Using wheel encoders and a gyrocompass, the microprocessor detected any motion and automatically corrected that motion to maintain position. In this mode of operation, the robot would automatically fight to stay in position, thereby creating a stationary platform from which to fire balls into the goal.

Autonomous target check was applied to continually monitor the camera output for the presence of the target light. If the robot was knocked off course and the camera lost the target, the ball firing mechanism halted and the robot automatically turned back in the direction of the light until the target was relocated. This feature ensured that balls would only be fired when they could be scored.

⇊ The inside of the frame includes a tray to support other robot systems. Dual motors on each side of the robot, along with a gear box capable of operating at different speeds, provide the robot with a range of power.

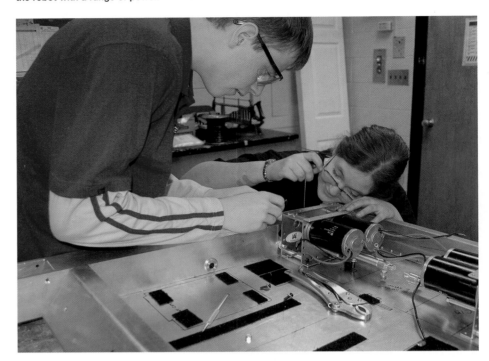

⇈ Tread tracks drive the robot and keep the center of the machine free to collect and lift balls.

The camera-aiming sequence was available in both autonomous and driver-control modes of operation. In driver-control mode, the robot panned the field, looking for the target light. Once the target was located, feedback was provided to the operator in the form of an illuminated green light emitting diode (LED) on the control board to signal that balls could be fired at the goal.

During prototype testing of the dual motor, four-wheeled shooter, the team detected that if the motors were operating at different speeds the ball trajectory would have a curve. The amount of spin imparted on the ball caused curved flight paths. They decided that varying the trajectory could be beneficial in match play. Since an independent motor powered each set of wheels, the system could be manipulated to produce the curved flight paths. A mini-joystick was added to the control board to give the operators the option to modify the ball spin and curve each shot.

⇄ Located at the highest possible location on the robot to improve accuracy and prevent blocking by other teams, the four-wheel shooter is supported by a truss. The support serves as a conveyor and transports balls from the base to the shooter.

## ⇄ Sequential Construction of Autonomous Actions

A series of program commands were written as building blocks to create autonomous routines. These commands accepted input from three types of sensors: encoders on each drive wheel to measure the wheel rotation, a gyroscope to measure the relative angle of the robot, and the camera for tracking the light above the goal. By creating simple commands that were strung together, autonomous programs were easily created for a variety of play scenarios.

To drive the robot forward a specific distance, a subroutine called "drive" was invoked. This subroutine specified the distance to be traveled that was then measured by the wheel encoders to determine wheel rotation. Output from the gyro was also used to keep the robot moving in a straight path. Any deviations from the path were detected and corrected by this subroutine. A control algorithm based on the travel distance was implemented to ensure that the robot reached the desired distance and did not overshoot its target.

Computer code named "turn" was written and executed to enable the robot to turn with tenths-of-degree accuracy. The gyroscope served as the sole sensor for this function, and a control algorithm monitored the turning rate and desired turn angle.

A delay routine was also included as a fundamental building block in the autonomous code. An internal counter measured the number of computer cycles that passed. Since the cycle time was constant, these counts were converted to a measure of time. The delay function was implemented to provide a time interval between robot actions, such as allowing for all motion to cease between turns, or to avoid collisions with pre-programmed functions executed by other robots.

The software subroutine "camera aim" relied on the output of the robot-mounted camera. The camera determined the location of the target light within the camera's field of view, and this data was used to align the robot base with the goal. Based on the camera output, the control algorithm determined the power required by each drive motor to accurately position the robot base in line with the goal.

A "base lock" command kept the robot in a stationary position when firing balls at the goal. Software parameters were set such that the robot would react suddenly and powerfully to disturbances, thereby keeping position despite any blocking attempts by opposing robots.

Additional software building blocks were created to shift gear speeds, to execute the ball-launching sequence, and to operate the conveyor, ball collector, and launcher functions of the robot. Thirty-two different autonomous modes were written and stored on the robot controller, with the desired mode selected using five two-way switches on the robot. Additionally, the start of the autonomous mode could be delayed by 1, 1.5, or 2 seconds to allow other robots to first execute their programs. Together, these options resulted in 128 possible autonomous play options—a variety that offered great utility to counter the expected opponent play.

## ⇄ An Incremental Plan

Team 1114's plan for software development was to make incremental improvements to a block of fundamental code that provided basic robot functions. The first iteration of autonomous control was added to drive the robot a fixed distance, and this mode was later enhanced with multiple loops running one after the other.

The next iteration introduced feedback control to keep the robot stationary. Upon verification of this function, a concise structure to add and remove sensors to monitor robot performance was added to the code.

With the drive system fully controlled, the software was extended to include commands to control the non-drive components of the robot, such as the launcher, conveyor, and ball collector. By adding features one at a time, the small block of fundamental code matured into a well-tested and effective collection of software that controlled every aspect of the robot.

⬇⬇ Polycarbonate panels define the ball-storage hopper and serve as a backstop for human player shots into the robot.

⬆⬆ The conveyor transports balls waiting to be shot out of the robot. The balls can be rapidly delivered from the hopper to the shooter, and the dual motors provide ample power to shoot balls as soon as they are delivered.

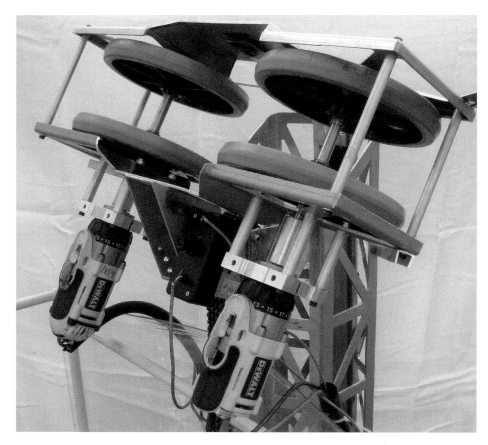

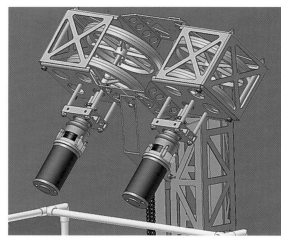

⇄ ⇈ ⇊ The similarity of the three-dimensional computer model and the final robot configuration are striking. The shooting system is solidly constructed to absorb vibrations and support the cantilevered design.

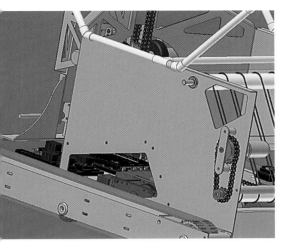

⬆⬆ ➡ Treads provide high amounts of torque to push other robots, to remain in a fixed location, or to climb the ramp, depending on what is needed at the time.

⬇⬇ The truss is firmly anchored to the base to resist the forces resulting from the large mass at the top of the robot. In addition to supporting the shooter and conveying balls to the top of the robot, the truss protects the electrical cables that power the shooting system.

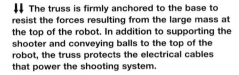

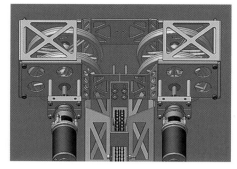

⇄ ⬆⬆ Electronics in the base of the robot are in plain view and protected from damage by the polycarbonate ball hopper.

⬇⬇ The robot control system was planned and constructed to enable functions to be easily executed.

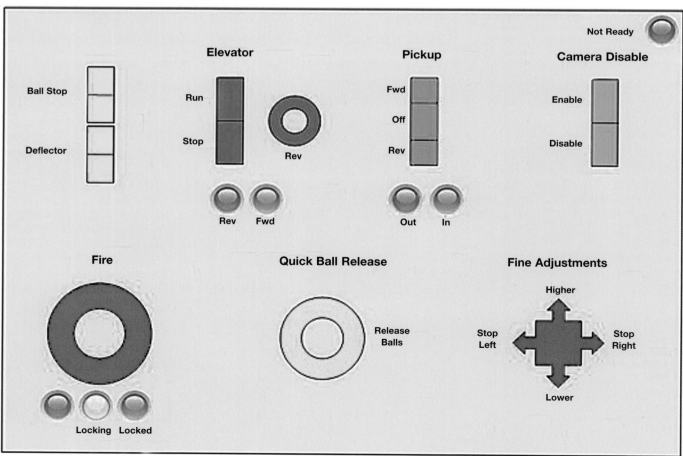

Not Ready

**Elevator**

Ball Stop

Deflector

Run

Stop

Rev

Rev    Fwd

**Pickup**

Fwd

Off

Rev

Out    In

**Camera Disable**

Enable

Disable

**Fire**

Locking   Locked

**Quick Ball Release**

Release
Balls

**Fine Adjustments**

Higher

Stop
Left

Stop
Right

Lower

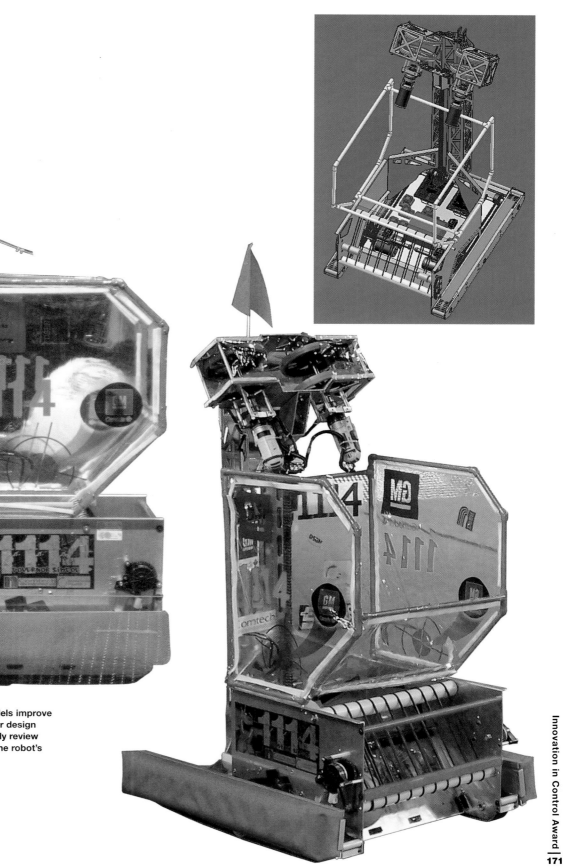

⇈ ⇉ Highly accurate computer models improve robot performance. The models foster design integration, provide a means to closely review fit and function, and are a record of the robot's construction.

# BUILDING BLOCKS FOR CONSTRUCTION AND CODE WRITING

## STEPPING THROUGH THE PROTOTYPE PROCESS

Three factors, as determined by Team 1629 from McHenry, Maryland, illustrate the connection between the machine and its operation: the physical robot, the software that runs the robot, and the control board that serves as the interface between the drivers and the robot. The team's success in each of these areas was the result of careful thought and coordination.

Student members of the team guided the design process. One group focused on robot structure to pick up, store, and dump balls, while a second group took on the task of designing a ball-shooting assembly. The robot structure subteam began with hand sketches and progressed to a full-size mock-up constructed with corrugated cardboard. The cardboard was easy to work with and provided a model of the ball-gathering and dumping system that could be examined and evaluated.

The second group designed and constructed a preliminary shooting system. The angle of shots and the spacing between the shooting wheels could be varied in this first design. The shooting frame and shooting motors were fastened with C-clamps to allow for easy modifications. These variations allowed for experimentation to determine the optimal parameters to propel the ball at the needed velocity while remaining within the design space allotted by the structure team.

Corrugated cardboard is a useful material to explore ideas and concepts. Because of its rigidity, the material can be used to construct full-size mock-ups to look at design possibilities.

Tests are conducted to determine the spacing between the shooting wheels and the best angle for the shooter. The launch velocity is a factor of the spacing between wheels, and the range is heavily dependent on the launch angle.

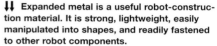

The ball collector powers a conveyor that lifts balls to the top of the robot and stores them in a magazine for later use. As with the shooter, the spacing between the conveyor and the backing plate is an important parameter to allow transport and prevent slipping.

Expanded metal is a useful robot-construction material. It is strong, lightweight, easily manipulated into shapes, and readily fastened to other robot components.

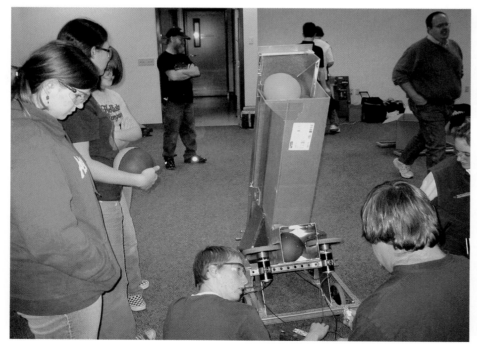

Located near the base of the robot, the shooter receives balls from the magazine and propels them from the robot. The magazine keeps the balls in a single line to prevent them from jamming. Positioning the shooter at the base improves robot stability.

A ramp is included in the design to dump balls into the lower goal. When needed, the ramp is lowered and stored balls are ejected from the robot by running the ball collector in reverse.

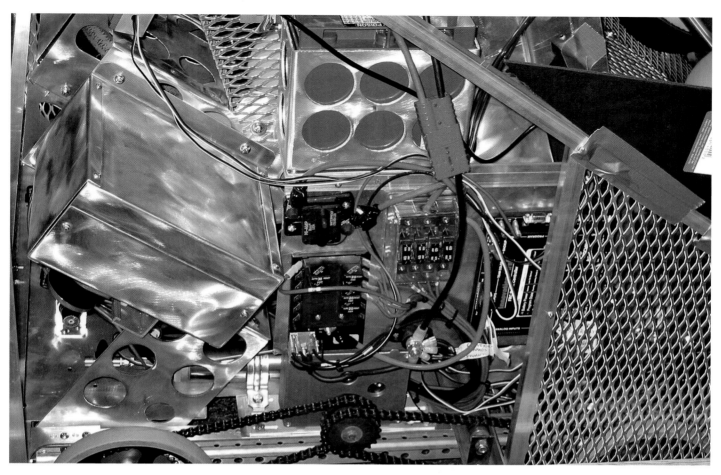

↑↑ ⇉ The electrical distribution system is arranged to be accessible yet protected. The robot controller and fuses are mounted on one side of the robot, and the voltage relays and controllers are mounted on the opposite side. Labels mark all wires to ease troubleshooting.

↑↑ Expanded metal provides a shield around critical components and contains balls within the robot. The robot, with its ability to score in both goals and climb the ramp, satisfies the team's original performance requirements.

Based on the findings of these two subteams, a robot base and shooting system was constructed and augmented with a conveyor to capture balls and lift them to the top of the robot. At that point, the balls exited the conveyor and entered a storage magazine, first constructed with cardboard, before being fed to the shooting wheels. In the final design, the cardboard magazine was replaced with expanded metal to contain the balls and provide visibility of the magazine contents. A ramp was also included on the final robot to dump balls into the lower goal. Careful and thoughtful wiring added to the robot's overall effectiveness. The resulting robot had a high degree of flexibility with an ability to score in both the high and low goals.

## ⇒ Controlling the Robot

Understanding computer code is often a matter of mastering the vocabulary that is used within a program. For the FIRST control system, the default program code assigns variable names that are pertinent to the generic control system. For example, a switch on the joystick connected to the third input channel on the robot controller was named "p3_sw_aux2" within the default software code that determined robot functions. This switch may have been associated with a motor relay instruction, such as one named "relay4_fwd," which operated the attached motor in the forward direction.

Software could be written to map the switch to the relay using the line of code: "relay4_fwd = p3_sw_aux2." In this example, the relay would be energized (and the connected motor would run) when the switch was pushed. However, the line of computer code was not the easiest to immediately understand.

Team 1629 instituted a simple yet effective strategy for making the hundreds of lines of computer code more understandable by defining new variables that were much more intuitive. With two lines of code, the arcane variables "p3_sw_aux2" and "relay4_fwd" were reassigned as the variables "dump balls switch" and "hopper dump relay." The mapping between the switch and motor became the more readily understandable line of code "hopper dump relay = dump balls switch"—a much simpler set of instructions.

By renaming each of the default variables with names that represented robot functions, computer code was written with vocabulary that was easier to understand. Replacing variables such as "pwm05" and "p3_sw_aux1" with "conveyor drive" and "fire ball switch" allowed code to be written with descriptive words that fostered understanding of the tasks being controlled. The new vocabulary resulted in concise statements for controlling the robot and minimized program errors.

A similar simplification process was applied in other control areas. For example, a switch was programmed to change the drive orientation to one that would make driving more intuitive. Because the robot had the shooter in the front and the hopper in the rear, it was necessary for either side to be able to serve as the vehicle's front, depending on the operation being performed. To avoid having to operate the vehicle "in reverse," a switch on the control panel was activated to designate which end of the robot would be the front. With the switch activated, the robot would always turn to the right when the right joystick was positioned in that direction.

Simplicity was also an advantage in Team 1629's algorithm for aiming the robot using the camera. By fixing the camera to the robot's central axis above the shooter, the camera output for the target light served as a direct input to the propulsion control system. The difference between the location of the target light and the robot direction indicated the amount the robot needed to turn to align with the goal. The status of the aim, measured as near, far, and centered, was displayed by LEDs on the driver control panel. With a click of a button, the driver could signal the robot to automatically adjust itself and lock onto the target.

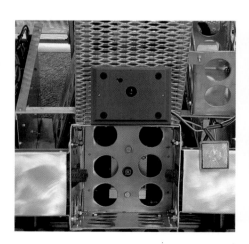

⇈ A camera is rigidly mounted to the robot right above the shooter. Easy-to-understand computer code maps the camera output to the robot's drive motors to aim the robot at the goal.

⇈ The control board alerts the drivers to the camera's actions and signals when the target has been located and the robot is locked on the target. Displays also provide the status of the robot aim before the camera is locked on the target.

## ⇄ Switching Modes of Autonomous Play

Because Team 1629 decided early that it would be critical to win the autonomous mode of each match, team members ensured that their robot was effective during this portion of play. To be successful in a variety of game alternatives, it was important to quickly change autonomous modes, depending on the capabilities of alliance partners and to respond to threats from opposing teams.

A clever selection switch was created and mounted on the robot controller to select the specific autonomous mode program for each match. The eight settings on the switch could be positioned to indicate any of 256 different autonomous operations. One group of switches specified the autonomous function (shoot, dump, or block) while another group of switches specified options within these three modes.

The autonomous mode switches provided a means to specify a large number of robot functions. For example, when shooting, the team could specify the number of balls to be shot and the number that would be saved for the offensive round. With an option to delay actions, a strategy could be immediately

crafted to avoid an opposing robot trying to block the team. If the opposing alliance had a plan to follow a straight line and create a moving obstacle, Team 1629 needed only to enter a delay option to allow that robot to pass before executing the shooting algorithm.

A series of options were created for autonomous mode play, including shooting at the upper goal, dumping in the lower goal, and driving straight to block an opponent. The camera served as the sensor to drive the robot toward the upper goal. Once in position, the software executed a locked-on routine that kept the robot aimed at the goal, even in the face of disturbances by other robots.

Computer instructions were also written that used dead-reckoning to approach the lower goals. The amount of time the drive motors ran was a variable set using the autonomous mode switches. If the robot traveled too far and the gate was prevented from lowering by the field, a routine to back-up the robot and then lower the ramp was included in this set of instructions.

## ⇄ Controlling Operations with an Intuitive Control Station

The driver's control station was designed to be easy to use, with a single joystick controlling the robot drive system. While the pilot positioned the robot, the co-pilot operated the scoring mechanisms. A collection of dials, switches, and indicator lights were arranged to allow the co-pilot to monitor and effect robot functions. Light emitting diodes mounted on the control board alerted the operators when the robot was in the optimal scoring position and signaled when system problems occurred.

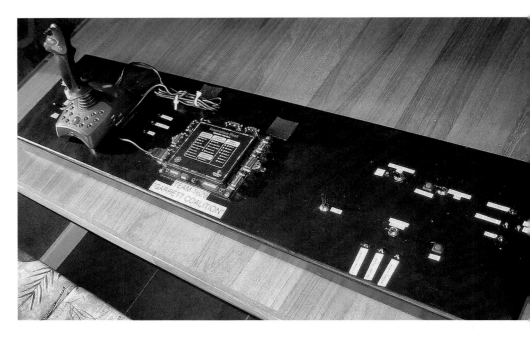

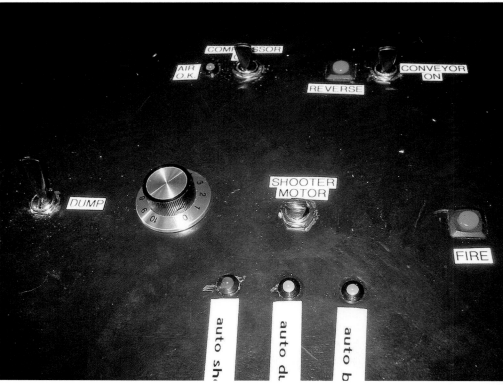

⇈ A single joystick drives the robot, leaving all other robot operations to the co-pilot.

# ⇄ Putting it All Together

To be competitive, the robot needed a sound mechanical structure, a robust and capable drive system and a control system that was effective in the autonomous and driver-controlled periods. Also, the robot and the operators needed to quickly adapt to changing conditions. FIRST Team 1629 met these challenges!

The team's many autonomous modes enabled them to win the autonomous play period in almost every match. The rugged design and high torque enabled the team to block other robots from scoring. Their automated target-lock and the ability to quickly switch between shooting and dumping provided Team 1629 with many opportunities to score. The robot would invariably make it back onto the platform at the end of the game. This well-coordinated team designed a great robot, easy-to-understand control programs, multiple modes of play, and an intuitive driver station—important building blocks that combined for overall success.

⬆⬆ The constructed robot consistently locates the target light and delivers balls into the goal. A low center of gravity benefits stability and keeps the robot on its wheels when climbing the ramp.

# SECTION 04

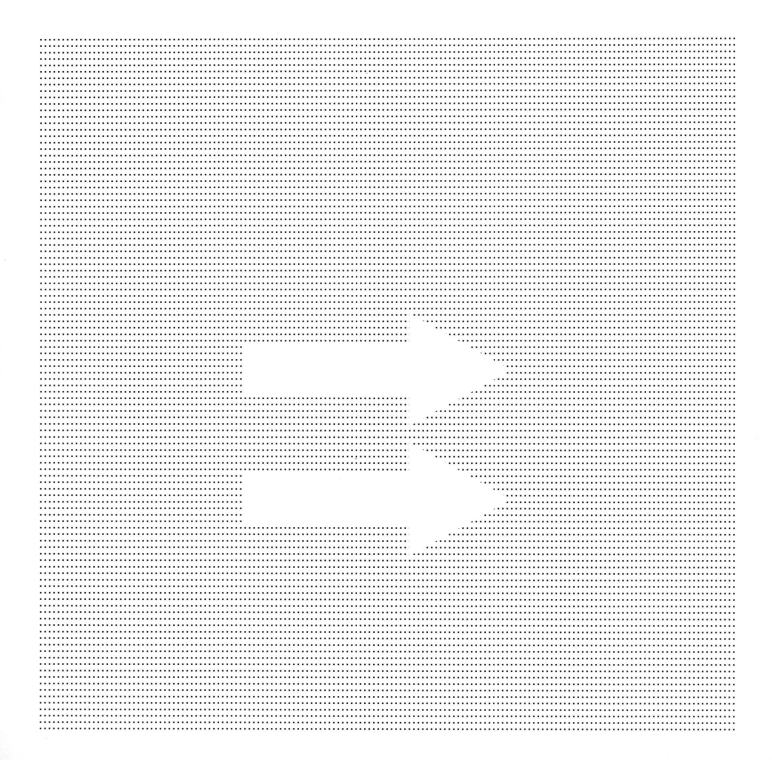

The Motorola Quality Award celebrates machine robustness in concept and fabrication. Winners of this award have strong, well-built robots that require little maintenance. Although the award criteria do not include on-field performance, winners of this award are typically among the highest ranked teams at the competition. As durable and reliable machines, the Motorola Quality Award-winning robots are effective competitors and valued partners.

Motorola is a Founding Sponsor of FIRST and has been extraordinarily generous. As a long-term supporter of FIRST teams and events, the company's commitment to workforce development has helped to inspire a whole new generation of scientists and engineers. The company has also funded independent evaluations of FIRST programs, and helped rebuild the FIRST LEGO League program in Louisiana and Mississippi in the aftermath of Hurricane Katrina. Top-level Motorola executives have served on FIRST's Board of Directors for many years, and a retired chairman of the Motorola Board currently serves as an honorary director of FIRST.

# Motorola Quality Award

# SCORING ON THREE WHEELS

## RESEARCH: THE FIRST STEP IN ROBOT DESIGN

A disciplined design effort begins with thorough research. Team 16, the Baxter Bomb Squad from Mountain Home, Arkansas, combined research with creative thinking to produce a winning robot named Two-Minute Warning.

After an initial brainstorming session to identify the robot's performance attributes, the team researched projectile motion and robot drive trains. This research promoted creativity and increased the team's awareness of the underlying concepts for each component. For example, the projectile motion study helped the team understand the dynamics of accelerating a stationary object to launch speed. With this information, the team determined the amount of power needed to launch balls into the upper goal. Similarly, research on right-angle drive trains—where the driving motor is mounted perpendicular to the wheel's drive shaft—prompted the team to consider bevel gears for the drive system.

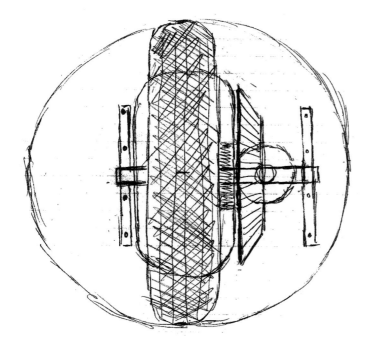

⇄ Bevel gears allow the drive motor to be mounted perpendicular to the drive shaft. A larger beveled gear on the drive shaft reduces the output motor speed and increases the produced torque.

$$T = (V_0 \sin \varphi)/g + \sqrt{[2(h - \tfrac{3.5 ft}{})]}/g$$

$$h = (39 ft^2 \sin^2 30°)/[2(32.2 ft/s)]$$

$$(39^2 \cdot .25)$$

$$/ 64.4$$

$$1521 / 64.4$$

$$23.618 \, ft$$

φ = 30°
$V_0$ = 39 ft/s
b = 3.5 ft
h = 23.618 ft
T = 1.7235
R = 60.49 ft

⇈ Understanding the physics behind a problem is an important first step before solutions can be generated. For propelling balls into a high goal, the team members must calculate the time, distance, and height of the ball as a function of launch angle and velocity.

# ⇄ Computer Modeling for High Performance

A methodical process of research, sketching, computer-aided design, manufacturing, and assembly contributed to the creation of this award-winning robot. Robot components were tested with prototypes, and the results were applied to improve the performance of each subsystem. Improvements were incorporated into the refined version of each subsystem and modeled with Autodesk Inventor computer-aided design software.

The computer models were very detailed and provided the team with a good idea of what the constructed robot would look like. In addition, the models became a tool to inspect each subsystem and identify areas for improvement. After the models were reviewed and approved, manufacturing drawings were created and forwarded to the machine shop for fabrication. A close review of the computer model ensured that parts would work as designed, eliminating a more costly trial-and-error evaluation process.

Once the parts were manufactured, the subsystems were constructed. Each component was tested and optimized before being integrated into the actual robot. The subsystem testing sequence identified problematic performance areas, and any needed improvements were made. Testing each system before it was installed on the robot reduced delays that might have otherwise occurred if each one had to be tested and improved after being installed on the robot.

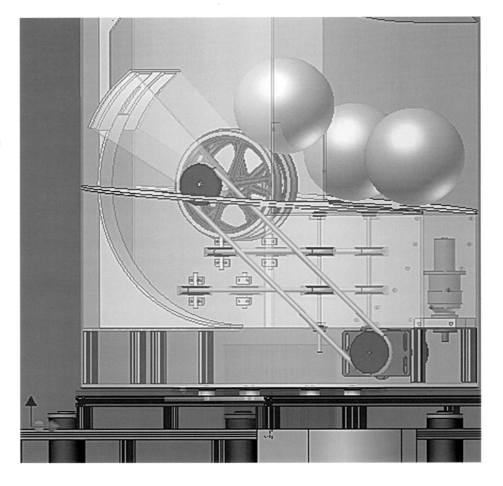

⇄ Three-dimensional views of the robot's systems illustrate how individual subsystems are related to each other. The image can be viewed on the computer from any perspective, enabling a complete review of the proposed robot.

## ⇄ Inspired by Bowling

Two features of Two-Minute Warning are especially interesting: the shooting device and the drive train. The team's initial experimentation with shooting mechanisms included a review of sports-based methods to mechanically propel a ball. A kicking mechanism was first explored. The device consisted of a swinging pendulum that contacted the ball and launched it into the goal. Testing indicated that this propulsion method was not effective because the foam ball absorbed a significant amount of the impact energy and, as a result, the ball had limited projection distance.

A baseball-pitching machine was investigated next. In this experiment, a rotating disk compressed a ball against a flat plate and shot the ball out of the mechanism. A review of this technique revealed that the single-wheel design yielded unpredictable flight patterns because of the inability to center the ball on the wheel.

Bowling was the next sport to stimulate ideas. The team realized that most bowling-ball return systems use a single wheel to propel the ball, but the wheel does not have a flat outer surface. A V-shaped groove around the circumference of a bowling ball return wheel centers each ball on the wheel and prevents misalignment.

Initial tests with a single grooved wheel were not successful. The wheel tread separated from the hub when rotating at the high speeds required for launching the ball. Rather than using a single grooved wheel, Team 16 used two wheels, separated by a small space, to propel the balls. This concept proved to be a reliable method for shooting.

The idea of using space as an alignment tool was also applied in the dual tracks of tube that guided the balls to the shooting wheel. Solid core tubing was cut to size, threaded on pulleys, and thermally joined to create a continuous belt.

Team 16 used polycarbonate, a strong, transparent material, as structural framing for the shooter and other robot parts. Large holes were drilled into the polycarbonate to reduce the robot's overall weight. In addition to being a strong structural material, the polycarbonate pieces added to the robot's elegance. The transparent pieces enabled team members to easily examine internal robot components.

⇄ Polycarbonate is a versatile construction material that offers strength and visibility. The material is easy to machine and is available in a variety of thicknesses.

⇊ Cords stretched between pulleys guide the balls into the shooting wheels. The space between the wheels centers the ball in the shooting device, and the curved back plate determines the ball exit angle.

# ⇄ Swerve Drive Propulsion

Two-Minute Warning's drive system was unique. The team's early research indicated that bevel gears could be used in a propulsion system in which the motor was mounted perpendicular to the wheel's drive shaft. The team envisioned a small beveled gear attached to the motor shaft and mated with the larger wheel-mounted bevel gear. This breakthrough allowed the team to realize its goal: designing a robot drive system that allowed the robot to transverse the field in any orientation.

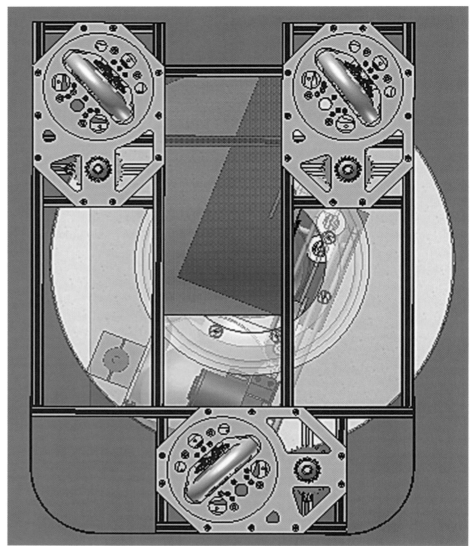

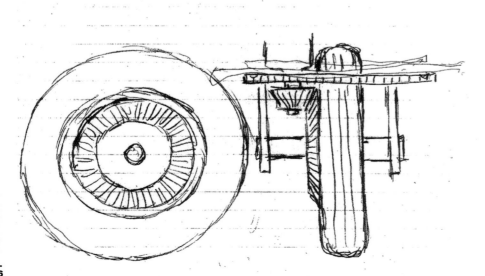

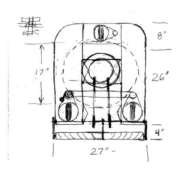

⇄ ⇅ Beveled gear transmissions, drive motors, and wheels are packaged into a composite unit. The small footprint of each propulsion unit enables the drive system to be distributed around the frame edges, keeping the center of the robot base free for ball collection and storage.

Advancing from hand-sketches to a computer model, the team designed a three-wheeled drive system. The orientation of each wheel module, consisting of the wheel, transmission, and motor, was independent of the other two wheels. The advantages of this system were the ability to independently control the direction of force delivered by each wheel and to decouple the relationship between the robot's orientation and its motion. Two wheels were located in the front of the robot, and the third wheel supported the load in the rear of the robot.

With this drive system, the robot could be facing the goal and move sideways across the field, swerving around obstacles, all the while not changing the robot's orientation. Acknowledging this performance capability, Team 16 called the mode of operation "swerve drive."

An advantage of the swerve drive system was the robot's ability to hold its position on the field. In pushing matches, Team 16 could rotate each wheel module to be perpendicular to the direction of the pushing robot. This orientation maximized the frictional force that any opposing robot would have to overcome before Two-Minute Warning could be moved.

Gear selection and control algorithms added to the effectiveness of this drive system. The propulsion system could be operated in high gear to achieve high speed, or low gear for maximum torque. When needed, the operators could alternate between these two conditions.

Computer algorithms made using the propulsion system intuitive. Each wheel's rotational speed and steering angle were monitored and regulated by the on-board robot control system. Using this data, the wheels were oriented in the correct position to drive the robot as commanded by the driver's joystick. If the joystick pointed to a specific direction, the robot moved in that direction.

The robot controller regulated the speed of each wheel during a turn, thereby producing the smoothest of turns. During turns, outer wheels rotated faster than inner wheels to prevent any wheel from slipping along the carpeted playing surface. The swerve drive system and sophisticated control algorithms produced a fast (18 feet per second [5.5 m/s]) and highly maneuverable robot.

⬇⬇ Three propulsion units drive the robot. Two are located at the front of the robot on either side of the ball-retrieval system, with the third balancing the rest of the robot. Each unit can be independently steered.

# ⇄ Squeezing Propulsion Components into a Small Space

Each wheel module was a complete propulsion system that included a motor, transmission, wheel, and steering mechanism. A drive motor powered each wheel. The vertical orientation of the motor created a compact design that did not require a large footprint and allowed the entire wheel module to rotate.

The transmission, positioned between the motor and the output bevel gear, reduced the high motor speed to propel the robot. A small servo motor was used in each transmission to shift between high and low speeds; the transmissions were standard hand-drill transmissions. An aluminum frame was manufactured to support the wheel and its power system.

A separate motor steered the wheel module. The steering motor drove a small sprocket, and a chain connected this sprocket to a larger sprocket mounted on the wheel frame. By energizing the steering motor, the entire wheel module could be rotated. A small white gear was also mounted to the steering motor output shaft. This gear triggered a rotational sensor to measure the wheel module's orientation – information that was passed along to the robot control system to coordinate motion.

↑↑ Each drive unit consists of a motor, transmission, shifting mechanism, feedback sensor, wheel, supporting frame, and steering mechanism. All components are mounted on a common frame, and that frame is rotated to point the unit in any direction.

↑↑ The yellow transmission allows the motor to be operated in either low or high speed. The transmission is shifted using the small black servo motor mounted at the base of the transmission.

↑↑ With two motors up front and one in the rear, side-to-side balance is a concern. Small rollers located near the rear of the robot keep contact with the playing field surface and prevent tipping.

# ⇄ Strong Results from Strong Design

The swerve drive system allowed the robot to be extremely effective at offense and defense. Because the orientation of the robot was independent of its direction, the robot was effective at collecting balls from the floor and shooting them into the goal. The robot's ability to transverse the entire field in four seconds, coupled with an ability to shoot balls from any angle, made Team 16 one to reckon with on the field. On defense, the team was equally skilled because of the independently steered wheels, which could be positioned to withstand the largest possible force delivered by opposing robots.

The modularity of the propulsion system proved to be a valuable asset during competition. One exciting match ended with another robot impaled in one of Two Minute Warning's drive wheels. The entire wheel module was easily removed and a damaged gear train replaced during a short time-out period enabling the machine to compete in the next match.

Careful planning benefited the team by saving time and resources. Since each design component was initially created using computer-aided design tools, the parts could be examined, evaluated, and verified before being manufactured.This planning and review process eliminated design mistakes and enabled the team to package a number of advantageous features into a compact space. Research, creative thinking, and clever design all were important attributes that produced the three-wheeled champion robot.

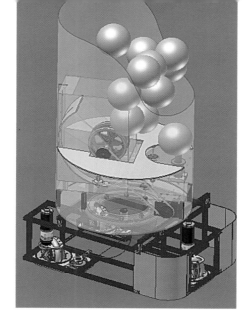

⇄ A see-through cylinder contains the balls waiting to be scored. The cylinder, with the shooter mounted inside, rotates to align itself with the goal.

⇊ With its pivoting three-wheel power system, dual-roller ball gatherer, internal conveyor system for ball transport, linear ball storage helix, and camera-aimed shooting system, the robot is an integrated system of systems.

# ONE ROBUST WO-BOT

## BUILT TO BE TOUGH

While other FIRST teams build robots, FIRST Team 141 from West Ottawa, Michigan Public Schools annually builds a WO-Bot. With participation in FIRST Robotics Competitions dating back to 1999, the team has many years of experience to draw upon. Based on the expected intensity of the 2006 FIRST Robotics Competition, the team decided that a strong drive system and a solid robot base would be essential. Experience served them well, as the team designed a robot capable of delivering and receiving aggressive play on the field.

Quality is a measure of a system's ability to meet design goals. A robust robot is one that is well designed and constructed to operate match after match. Over the course of the season, Team 141 played 46 matches and never had a component fail. By anticipating the on-field action, Team 141 designed a robot to maximize performance and minimize maintenance.

### ⇄ Developing a Design Plan

The team's on-field success began with a structured approach for decision-making. An analysis process known as the Pugh method was used to examine the competition and develop initial plans for the robot.

The Pugh method was applied to evaluate design choices against performance criteria. Used with three or more alternative choices, the Pugh method designates one choice as a benchmark and then compares every other choice to that standard for each design criteria. The value of the criteria can be weighted to emphasize the importance of some criteria in comparison to other design factors.

Design alternatives for shooting, propulsion, ball pick up, and storage were evaluated against a list of 40 design attributes. One design was identified as the baseline option, and other alternatives were compared against the baseline to determine if the alternative was better than, the same as, or not as good as the option. A point value was assigned to each evaluated option (1, 0 or –1) and multiplied by a weighting factor for each design attribute. The sum of the scores for each alternative identified the optimal design.

As an example, three driving methods were examined: a four-wheel drive system where each wheel had its own motor, two wheels that could be steered (much like a car), and a design that used retractable wheels. Some of the important design attributes included serviceability, ability to be manufactured in four weeks, traction, and maneuverability. The Pugh analysis guided the team to a drive system that maximized pushing power when going straight and used retractable wheels for increased maneuverability during turns.

In addition to using the Pugh method to make design choices, the weighted design attributes were a declaration of principles that helped keep the team focused throughout the design and construction process. A team member began to stray from the design guidelines, could be reminded of the established design requirements and so realign that creativity to honor the team's ranked list of design attributes.

Team 141 also applied this structured decision-making method during the competition when teams selected partners for the final rounds of play. A list of team and robot attributes was identified and potential alliance partners were evaluated based on their predicted performance in each category. With this decision-making methodology, Team

141 strategically evaluated all factors and selected the optimum partners to create a winning alliance.

## ⇄ Evolutionary Design Improvements

A series of tests and reviews advanced the robot's performance. Balls were collected from the floor and lifted to the top of the robot, where they were shot or stored. Large-diameter wheels provided clearance to collect balls at the front of the robot, with a set of brushes placed in front of the wheels to prevent the robot from running over any balls. Once under the robot, the balls were guided by a set of rails to an elevator that lifted the balls from the floor.

**↑↑ ↓↓ Large, pneumatic wheels provide ample clearance for a frame that doesn't interfere with ball collection. Brushes in front of the wheels deflect the balls away and prevent the wheels from running over balls and possibly interfering with the drive system.**

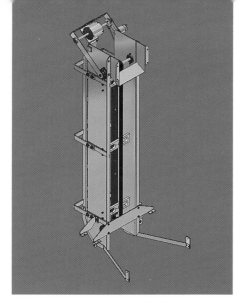

⇄ The single-belt conveyor is for ball transport and storage. The shooting wheel is mounted at the top of the elevator, providing a convenient storage system for balls waiting to be shot.

The first iteration of the design cycle included an elevator that included 24 rollers. Sets of rollers were connected with an elastic cord that wrapped around a center rod, and a single motor drove the center rod to power all 24 rollers. Testing revealed the fragility of a system with so many moving parts, and the elevator mechanism was simplified to be a single belt.

Although the team used a shooting wheel to propel the balls, this was not the only system considered. The shooter was an evolutionary design that began as a catapult mechanism.

The catapult proved to be too heavy, and was replaced by a small-diameter, rubber-coated pulley that was fixed to a spinning shaft. Because of the wheel's small diameter, the shooter wheel had to be rotated at a very high speed to effectively propel balls from the robot. The trajectory angle of the ball was fixed, so the robot could only score from the top of the ramp. While limited in scoring positions, the close range yielded a high degree of accuracy and offered natural protection from opposing robots.

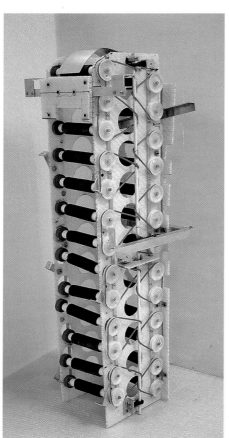

↑↑ An earlier version of the elevator was much more complicated and included 24 separate rollers all powered by a common drive shaft. The single-belt elevator was selected over the roller system because of its simplicity and increased reliability.

↑↑ A spare drive cord is stored on the shooting wheel's shaft, ready to be used if the installed cord snaps. Should the drive cord fail, the spare can be installed without disassembling the shooting system.

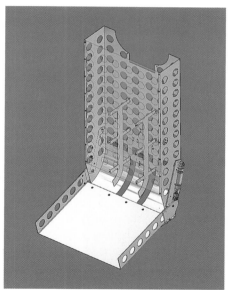

⇄ ⇊ To deliver balls into the lower goal, a ramp lowers from the front of the robot and three columns of balls roll into the goal. Because the balls are stored in an orderly fashion they do not bind on one another and easily roll out the ramp.

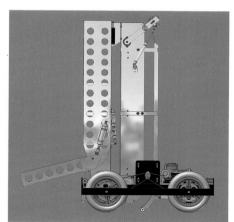

⇈ Gathered balls are transported up the elevator to be either shot or stored. The storage hopper at the front of the robot is capable of holding a large number of balls.

The front of the robot included a storage bin to hold balls that were not immediately shot from the robot. Scoring in the lower goals was made possible by lowering a ramp at the front of the robot. When the gate was lowered, the three columns of stored balls simply rolled out of the robot and into the lower goal. Separating the balls into three columns prevented them from binding on one another and jamming the gravity-fed scoring mechanism.

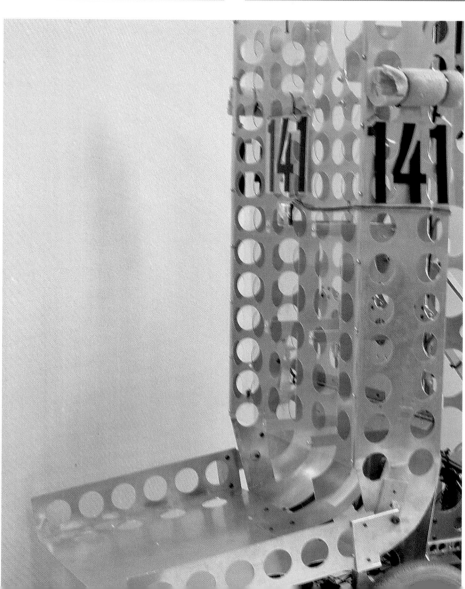

## ⇌ Clever Power Transmission

Power was transferred from the drive motors to four large-diameter pneumatic wheels using timing belts rather than chains. One timing belt connected the drive motor to the transmission, and two other belts connected the transmission to each wheel. Belts were also used in every other application on the robot, including the ball-gathering roller, elevator and shooting wheel. Over many years of experience in the FIRST Robotics Competition, Team 141 has found this transmission technique to be effective and easy to maintain.

⇌ Pneumatic pistons at the rear of the robot are lowered automatically during turns to eliminate friction from the rear wheels. With the rear wheels raised, the robot pivots on the front wheels to turn.

⥥ Belts connect the output of the drive motors to the wheels. The motors are mounted on the outboard side of the robot, an orientation that makes use of open space over the wheels and frees valuable real estate inside the robot base.

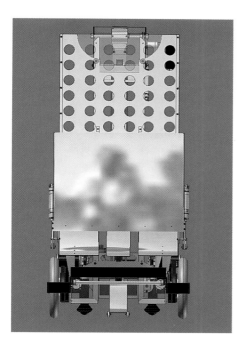

↑↑ Low-friction material is used on the turning piston ends to reduce drag. In the retracted position, the pistons ride just above the playing surface, allowing for quick deployment when needed.

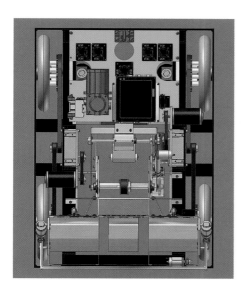

↑↑ Located between the rear wheels, the electronics are shielded from damage but remain accessible. Because of the high demand for compressed air to operate the turning wheels and front ramp, a compressor is carried on the robot.

The large-diameter pneumatic wheels were selected to provide the needed ground clearance to pick up balls and maximize torque. The large wheels provided room to mount large diameter sprockets on the wheel hubs, thereby reducing the speed and increasing the torque of each wheel.

The rubber wheels and carpeted playing surface resulted in a high amount of friction. This was an advantage that enabled Team 141 to hold ground when shooting and when playing defense. However the friction was a disadvantage during turns, as each of the four wheels gripped the carpet and would not easily slide. To counter this opposing force, the robot was designed with a set of caster wheels that would automatically be lowered during turns. These wheels were located in line with the robot's rear drive wheels. When deployed, these wheels greatly reduced the friction force during turns and enhanced maneuverability.

Team 141 relied on pneumatic power to deliver linear force throughout the robot. Pneumatic systems are easy to install and have high reliability. The caster wheels were positioned by pneumatic pistons, as were the elevator's ball diverter and lower-goal scoring ramp.

The pneumatic compressor and air storage cylinders, as well as the robot battery, were located over the rear drive wheels. This location maximized the load placed on the rear wheels and increased the effectiveness of the wheels to transmit torque. By being mounted at the lowest possible locations in the robot, these components also improved the stability of the robot—an important condition that prevented tipping during pushing matches and when climbing the ramp.

## ⇉ The Robust WO-Bot

Team 141 capitalized on its experience competing in the FIRST Robotics Competition to produce a very durable robot. A structured design methodology not only identified preferable robot systems but also produced a list of attributes to guide the team's creative design process. The team showed maturity in its ability to modify developed systems, such as the roller-driven elevator and catapult, which did not initially perform well. Combined, the design foundation and improvements greatly improved the team's overall performance.

The team's goals were to build a robot that scored points, operated consistently, and required little maintenance. Aided by their structured decision-making procedure throughout the design process, the team built a robot that was strong, reliable, and effective. While other teams might simply claim they built a robust robot, this team from West Ottawa public schools took tremendous pride in the creation of their "Robust WO-Bot."

# KEEPING IT SIMPLE

## SIMPLICITY AS A DESIGN PHILOSOPHY

KISS IT may seem like an unlikely plan for building a competitive robot, but that was the winning theme for Team 191, from Rochester, New York. With 15 years of FIRST experience, the team known as the X-Cats created a champion robot based on the principle "Keep It Super Simple in Technology."

The KISS IT principle prompted the team to choose simple, robust designs over complex and potentially more problematic ones. The team began their design process by examining the robots built by the X-Cats for each FIRST Robotics Competition since 1992. After a thorough review of 14 different robots, the team identified winning strategies and successful subcomponent systems.

This benchmarking review recalled successful X-Cat strategies such as the utility of proven design concepts, determining the "killer application" for a given year's robot competition, and leaving plenty of time for practice before a competition. The review also reminded the team to avoid problematic strategies such as developing too many new technologies during the build period, building a robot with a high center of gravity, or creating a machine that attempted to do everything.

During the review to guide the 2006 robot design process, proven subcomponent designs from previous robots were recognized and presented to the entire team. As a result of this process, design models for each of the following systems were identified: robot frame and drive system, ball retrieval and storage, and shooting. These subsystems, each a proven success in an earlier robot, were integrated to create a robust and reliable robot for the 2006 FIRST Robotics Competition.

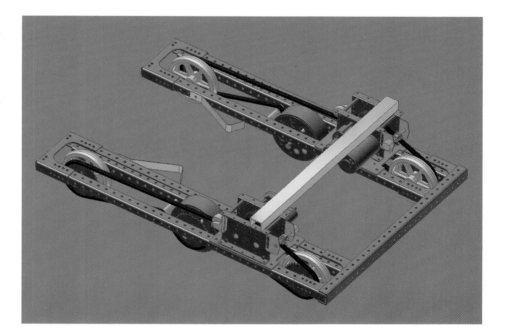

## ⇄ Robot Frame and Drive System

The 2006 X-Cats robot frame and drive system were based on the 2005 X-Cats design, which demonstrated high reliability and maneuverability. Because the 2005 design required no maintenance during the entire season of competition, it was a logical choice as the starting point for the new robot. The frame was constructed with structural aluminum supplied in the 2006 FIRST Kit of Parts and off-the-shelf extruded aluminum. The frame was easy to assemble using standard bolts as the fasteners. The predrilled structural aluminum allowed for quick modifications when each subsystem was added to the robot. The strong frame absorbed the shock from collisions and protected the robot's more delicate systems, such as the electronics and the ball shooter.

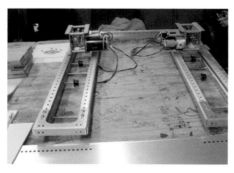

⇄ ⇈ A six-wheel drive system mounted on the standard frame provides ample torque for the 2006 competition. The design keeps at least four wheels in contact with the ramp and assists the robot's ability to climb on the platform.

↓↓ Off-the-shelf transmissions capable of shifting between low and high gear are used in the drive system. A single chain drives all three motors on each side of the robot, with idling gears mounted to keep sufficient contact between the chain and sprockets.

The drive system was the epitome of the KISS IT principle because it was constructed from material supplied in the FIRST Kit of Parts and off-the-shelf parts that required little machining. A two-speed transmission was purchased from a vendor and modified to provide higher torque and optimal speeds of 3.5 feet per second (1.1 m/s) and 9 feet per second (2.7 m/s). Two motors were used in each transmission to obtain high levels of power and to reduce the load on each motor. This drive system had a low center of gravity—an important feature for a stable robot that needed to climb ramps.

Each side of the robot had three wheels driven by the transmission, with the middle wheel slightly offset from the outer wheels. The offset wheel enabled the robot to ride on only four wheels at a time. This design allowed the robot to slightly rock when the momentum shifted from the front set of wheels while moving forward to the back set of wheels when in reverse. Compared to a design with all six wheels always in contact with the ground, the offset-center-wheel drive system was easier to maneuver because the design decreased resistance during turns.

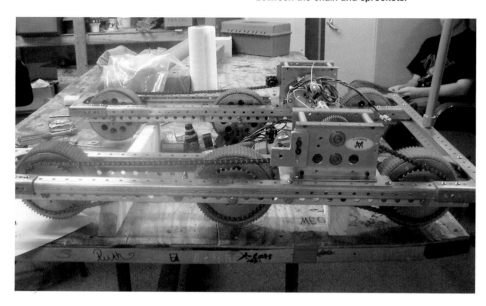

↑↑ A compact design, with the majority of heavy components mounted low in the robot, improves the overall stability of the robot. A compressor provides pneumatic power to switch the drive system speed.

## ⇄ Ball Retrieval and Storage

Twelve years ago, the X-Cats designed a ball-retrieval system that utilized a flat belt to guide the balls along a flat plate. This 1994 ball-retrieval design was simple, reliable, and quickly picked up balls off the floor. It was an easy decision to resurrect this subsystem design for the 2006 robot where one set of conveyor belts guided the balls into the robot, and a second set of belts pushed the balls up a vertical plate inside the robot. The conveyor belt system was constructed using timing belts with PVC spacers placed between the timing belt pulleys to keep the belts aligned. Both sets of belts were driven off of a common motor, thereby reducing the complexity of this subsystem.

⇊ Inspiration for the ball guide came from the 1996 X-Cats robot. In that design, a tie-dyed, U-shaped trough runs along both sides of the robot to move balls from the rear to the front of the robot.

⇈ The 1994 X-Cats robot served as the model for a ball-retrieval system. A flat belt compresses balls against the floor and lifts them into the robot.

⇊ Belts power the 2006 ball-retrieval and conveyor system. A motor powers the middle set of rollers, with the drive belts connected to the forward and upper rollers.

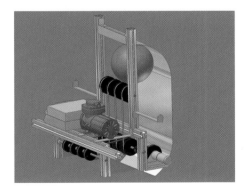

⇈ PVC spacers separate pulleys and ensure that the retrieval belts do not overlap and bind on each other.

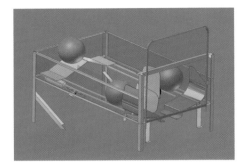

↓↓ An inverted, V-shaped deflector determines the path for each ball. Rails, gently sloped, guide the balls from the retrieval area to the launch mechanism.

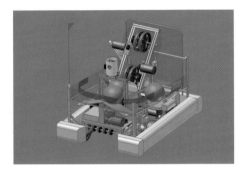

↓↓ A prototype was constructed and tested to determine the optimum angle for the ball ramp. The heights at the exit location of the ball-retrieval system and the entrance to the ball launcher were adjusted to create a gentle ramp that reliably advances balls.

Once picked up by the robot, the balls were stored and directed to the shooting mechanism in the rear of the robot. The 1996 X-Cat robot guided balls using a trough that ran along both sides of the robot. The angled trough used gravity to move the balls. This design inspired a similar system in the 2006 robot, where a set of rails directed the balls along the outer edges of the robot. An inverted-V ramp was located at the entrance and diverted the balls to each side of the robot. Prototype testing with a cardboard mock-up of the ball storage system determined the optimum angle for the storage ramp to be ten degrees.

At the end of the rail guides, the balls were restrained by a servo motor-controlled gate. When the gate was open, a single ball passed through to the loading mechanism, and all other balls would be prevented from interfering with the launching system. A hinged lever lifted the balls into the shooting system. A ring at the lever's end held the balls and a pneumatic piston at the opposite end of the lever provided the force to rotate the lever and load each ball into the shooting system.

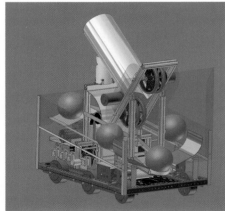

↑↑ A cantilevered arm lifts each ball into the shooting wheel. The arm acts as the trigger for the launch system.

# ⇄ Shooting Mechanism

The 2002 X-Cat robot served as the benchmark platform to explore shooting technologies. This review was combined with projectile motion physics and a study of commercial baseball pitching machines to understand the positive and negative qualities of shooting systems. The most promising shooting method was a pair of spinning disks through which the ball passed. The ball would be slightly compressed as it passed through the narrowest section of the spinning disks and in turn be shot out from the disks once the opening grew larger. Concepts were mocked up as stand-alone fixtures to examine the functionality of possible design alternatives.

The first concept used a single motor and a solid shaft to fire the ball, but the motor loads were too high. A series of experiments led to the final design: two sets of wheels driven by a single motor. Two factors surfaced during this review: the ball alignment and the forces exerted by the wheels on the ball. The paired wheels were chosen because of their ability to consistently center the balls in the shooting mechanism. Testing revealed that optimal forces were delivered with one urethane wheel and one pneumatic wheel. This combination provided a strong grip on the balls and accommodated compliance and surface variations from one ball to another.

Two parameters were critical for the ball's trajectory: the rotational speed of the shooting wheels and the angle of the shooter. The angle of the shooter was critical because of its significant effect on the trajectory. Testing determined the relationship between the shooting wheel speed and the distance a ball traveled, as well as the optimal angle of exit for the balls. The frame of the shooter was fixed at a 53-degree angle, and the shooter frame was designed to allow for slight angular adjustments. The rotating speed of the shooting wheels was measured with potentiometers mated to the main drive shaft. The final design produced an accurate shooter that launched balls into the goal from nearly any field location.

⬇⬇ The speed of the shooting wheel is monitored and controlled. The green gears mounted on the outside of the shooting wheel frame are coupled with a potentiometer to measure rotational speed.

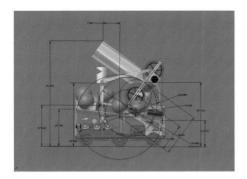

⬆⬆ Planning, analysis, and review ensure that the machine will operate as intended. With all objects in the CAD model defined, the distances between any two objects can be accurately determined.

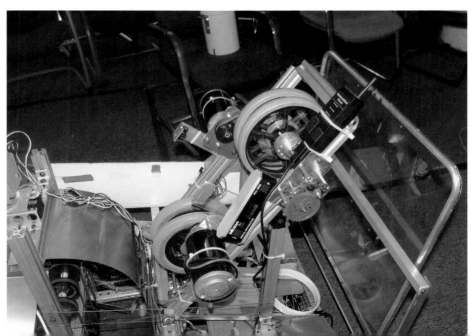

## ⇄ An Integrated System for Superior Performance

Once all designs and subsystems were constructed, a prototype robot was built and tested. CAD software was used in each step of the design process to model the robot and integrate individual design features. Planning in CAD was an efficient technique that saved time and documented the robot configuration. Integration problems were easily identified on the CAD models, and changes were made before the actual robot was constructed, saving time and resources.

The electronics were positioned in accessible areas, properly harnessed, and clearly labeled. Careful attention to layout and wiring, combined with clustering electronics near their functional systems, greatly aided in debugging and maintenance.

The KISS IT theme allowed the X-Cats to develop a prototype robot in a shorter period of time, thereby allowing more time to test, evaluate, and make modifications. The reliance on proven designs from past robots was a winning strategy that avoided mistakes and integrated the best features from a number of previous designs into a competitive robot. As demonstrated by their finish as one of the top teams at the Finger Lakes (New York) Regional FIRST Robotics Competition, the X-Cats know how to KISS IT and build a great robot.

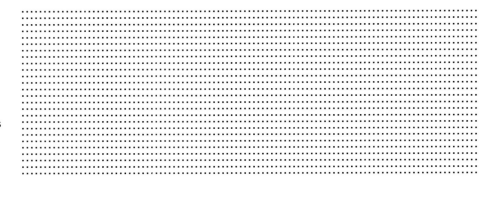

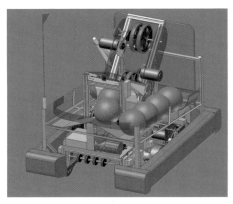

⇄ The sides of the robot store eight balls. An effective ball retrieval system is paired with a fast shooting system to rapidly gather and shoot balls.

⇅ Neatness counts when it comes to electrical layout and distribution. The compact nature of the design requires that all space be efficiently used.

There is much to learn from the past when it comes to robot design. The 2006 X-Cats robot is a combination of a number of previous innovations that proved to be effective on the competition field.

# ROBOT ART: GOOD LOOKS AND GREAT DESIGN

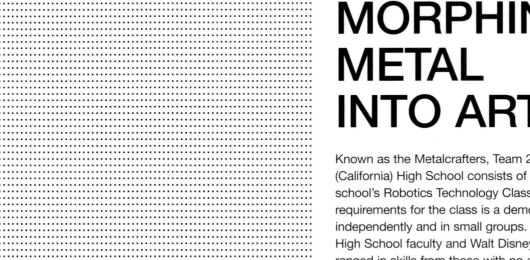

## MORPHING METAL INTO ART

Known as the Metalcrafters, Team 207 from Hawthorne (California) High School consists of students enrolled in the school's Robotics Technology Class. One of the entrance requirements for the class is a demonstrated ability to work independently and in small groups. Mentored by Hawthorne High School faculty and Walt Disney Imagineers, the students ranged in skills from those with no experience to ones with advanced experience as machinists. Working as a team, they were able to imagine an elegant solution and craft that solution into a work of art.

⇇ Team members brainstorm 2006 robot designs with Walt Disney Imagineering Mentor Michael Gordon (right).

⇊ An initial sketch of the robot includes a rotating turret and a distinctive S-shaped hopper. A hand sketch promotes discussion and records presented proposals.

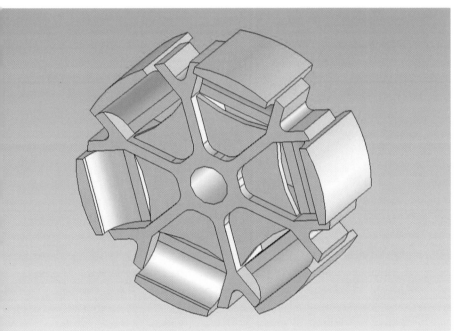

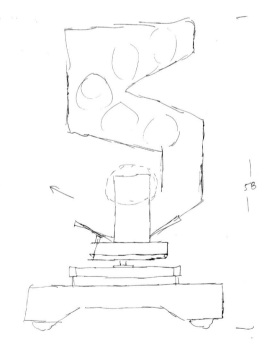

⇈ The rear wheels include articulating pads that rotate slightly to provide a constant source of traction and adapt to different field surfaces such as the ramp and platform.

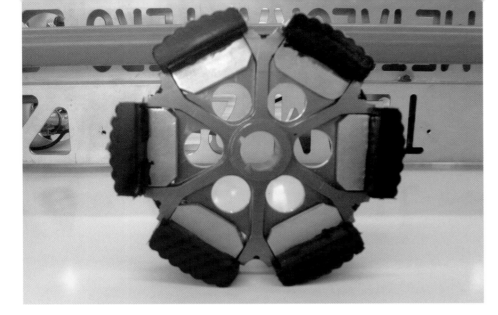

The team's work began with brain-storming sessions that produced simple sketches. From there, CAD models were created for all of the machine's components. No one task was left to an individual, but rather, each aspect of design, fabrication, and assembly came from a team effort. By working in groups, the team produced a robot that was incredibly robust and gorgeous.

The team's careful approach to design and its in-house fabrication is demonstrated in the robot's frame, drive train, and wheels. In each case, input from all team members was sought for original ideas before subteams proceeded to design, fabricate, and assemble the components. The subteams came together during the final assembly process where the subcomponents were joined together to create a beautiful machine.

⇈ Dual sets of articulated padded traction wheels provide constant contact with the playing field surface. The compressor is mounted in line with the drive train to help provide traction.

⇉ Layout of the pitching ball machine exit is crucial to achieve the desired launcher elevation and pinch points. These factors determine the launch velocity and trajectory.

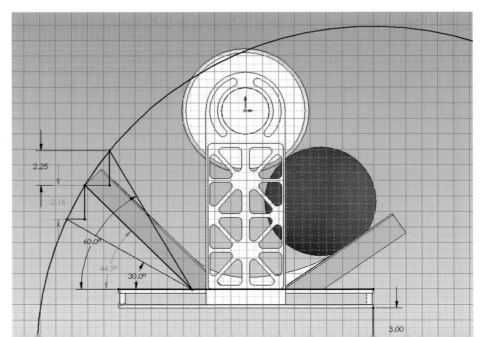

# ⇄ Building a Strong Frame

The robot frame was designed with aluminum box tubing that was welded together. Sections of each tube were removed to reduce weight yet preserve strength and rigidity. Students, guided by mentors, operated sophisticated machines and produced the attractive precision parts that were used to construct the frame. Following milling, the individual tubes were welded together to form a strong chassis for the rest of the robot to be built upon.

The base doubled as a bumper for the robot to protect the electronic components, motors, and wheels during the most severe collisions. The frame was much more than a simple box for it incorporated all the necessary supports and brackets to attach motors, bearings, and other pieces of hardware. A close working relationship between the chassis designers and other subsystem teams ensured that the frame supported all robot systems.

⇄ ⇈ The geometry of the chassis cross member is examined and verified prior to programming the Computer Numerical Control (CNC) milling machine. Once programmed, the CNC accurately cuts out patterns to minimize the component's weight.

⇊ The edges of the welded chassis are filed to eliminate sharp edges and prepare the chassis for anodizing.

order to minimize drag. This wheel arrangement offered power and mobility—factors the team predicted would be important in the competition.

Initial three-dimensional computer-aided design drawings of the wheels served as input files to define tool paths for a Computer Numerical Control (CNC) milling machine to follow. In addition to designing and manufacturing the wheels, the team also fabricated the drive system's sprockets using CNC machines. This capability greatly increased the team's ability to prototype and test ideas. By designing and machining the parts in-house, the team was able to optimize the designs for durability, weight, and cost.

The front omni wheels consisted of an aluminum hub with Delrin rollers to provide the needed strength and mobility, while the rear wheels were fabricated with custom-poured urethane pads. In addition to the wheels, the team also designed and machined the mold cavity to create the pads and then used the molds to produce the traction pads. Here, too, the in-house expertise of the team accelerated the design process.

The underbody of the chassis revealed the layout of the custom-designed wheels. The wheels were designed to maximize torque and to minimize friction during turns.

⬇⬇ CAD drawings serve as the input files to create tool path instructions for CNC milling. The in-house process saves time and cost.

⬆⬆ One-inch (2.5 cm) -thick aluminum bar stock is machined on a three-axis CNC milling machine to produce the hubs of the front omni-wheels. The omni-wheel rollers slide across the playing field surface and reduce friction during turns.

Both sets of wheels were mounted in the robot chassis using the pre-designed support structures for the wheels' axles and bearings. The frame also supported other robot components, such as the pneumatic compressor, pistons, and transmissions, in a close-packed, functional arrangement. Each wheel was paired with another wheel that was offset 60 degrees to keep a roller or pad in constant contact with the playing field surface.

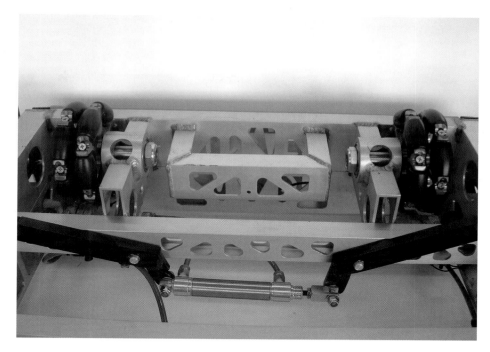

➡ The omni-wheels are mounted at the front of the robot. Two wheels are used on each side and clocked to provide continuous surface contact.

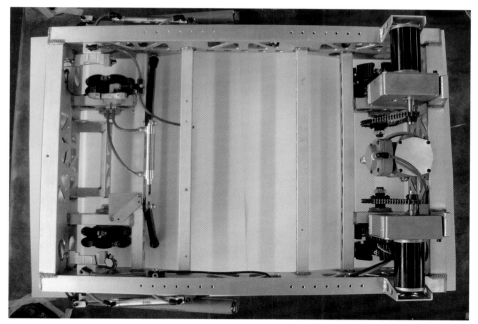

⬆⬆ The team's careful approach to design and its in-house fabrication is demonstrated in the robot's frame. In addition to the wheels, the frame also supported other robot components, such as the pneumatic compressor, pistons, and transmissions, in a close-packed, functional arrangement.

## ⇄ Assembling the Pieces

Additional craftsmanship was applied to create the turret-mounted shooting mechanism. Anodized aluminum plate was used as the material for many of these components—a technique that added great visual appeal to the assembled robot. CNC milling methods converted generic aluminum plate into elegant structural supports that were lightweight yet still strong.

A variety of techniques were used to join individual pieces of the robot together. Joints that formed foundations, such as the base of the turret, were welded. Subcomponent systems, such as the shooting system, were attached to the robot with threaded fasteners to facilitate removal should

repairs be necessary. The ball hopper was constructed from lightweight sheet metal, held in place with rivets. To maintain rigidity and simultaneously reduce weight, flanged holes were formed throughout the sides of the ball storage hopper.

The shooting device was designed to apply a progressive pinch to the ball between its entry and exit locations. Testing revealed that the progressive pinch was more effective than a single initial pinch. The hopper design narrowed at its base to sequence the balls and allow only a single ball at a time to enter the shooting system. The balls were centered in the shooter using an aluminum angle as a trough to guide the balls. A set of two adjoining wheels, with their hubs

shielded to reduce air friction, propelled the balls from the robot.

The heaviest parts of the robot, including the battery, air compressor, and drive motors, were positioned as low as possible in the robot to improve stability. By lowering the robot's center of gravity, the team maximized its ability to climb the ramp without tipping.

Careful attention was given to the layout of the robot's pneumatics and electrical systems. In each case, the components were organized into modular systems that could be moved on and off the robot for testing, maintenance, or troubleshooting. Using detailed wiring schematics and the labels on all wires and electronic components, the team could easily trace the system to detect any anomalies.

⇈ The hopper and turret are clamped in place while additional parts and designs are finished and fitted.

⇈ CNC machining on the turret base removes weight without sacrificing strength. Inserts are machined into the anodized plate to support components mounted on the plate.

⇈ The completed chassis, turret base, and pitching machine await the installation of wiring to complete the design and produce a working robot.

⇉ Rivets are used to assemble the anodize, aluminum ball hopper. Flaring the edges of the holes stiffens the thin plate and increases the structure's rigidity.

⇊ The clearance between the shooting wheel and guide rails is carefully designed to produce a progressive degree of pinch as the ball moves through the shooter.

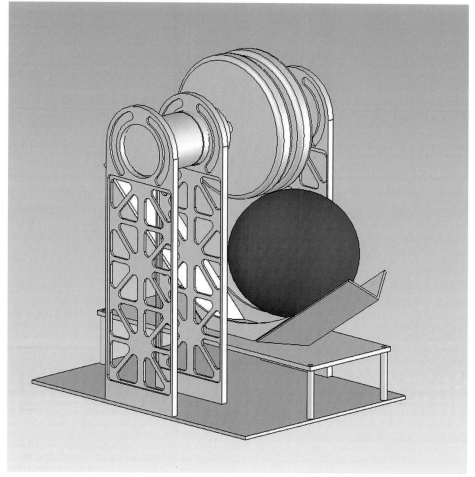

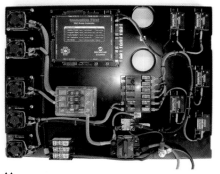

⇈ The electronic control board is designed to drop into the center of the chassis. The board controls all motors and actuators on the robot.

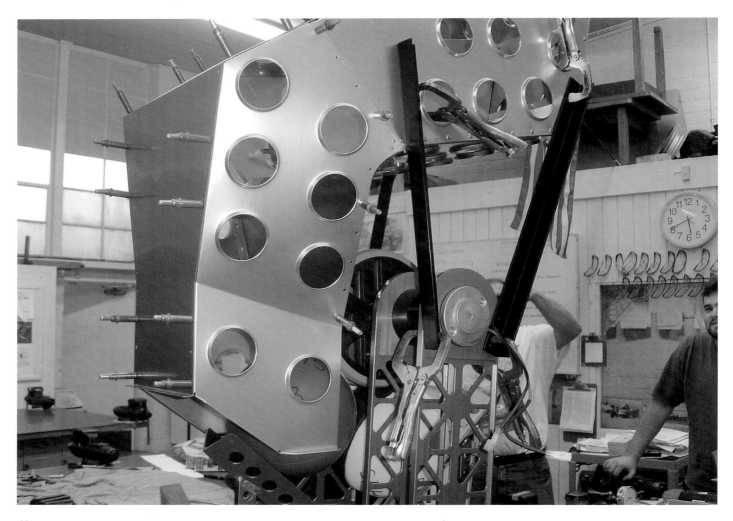

⬆⬆ Cleco clamps temporarily hold the aluminum sheet panels together. The clamps offer the freedom to test-fit pieces before permanently riveting them together.

## ⇄ Quality Robot and Quality Team

The assembled robot was strikingly similar to the preliminary design sketch. The latticed structure resulted in a machine that would be comfortable on the competition field or in a modern art museum. With its two-toned elements and geometric patterns in structural members, the robot had an open and airy appearance. In addition to this artistic flair, the design had functional relevance with its structural strength and visibility to internal system components.

The modular approach to subsystems worked well with the team's overall organization, where each subsystem was designed and fabricated by a separate group of students and mentors. Bringing the modules together on the final robot brought the team itself together. Just as the robot had many diverse parts that were assembled to form a robot, Team 207 itself had members with a diverse set of backgrounds what joined together to form a strong team. For both the robot and the team, the whole was greater than the sum of the individual parts.

⇒ The robot holds balls in its hopper while firing a shot from the front of the robot. All components of FIRST Team 207's robot are designed, machined, and fabricated in-house at Hawthorne High School.

# DOING ONE THING BETTER THAN ANYONE

## DOMINATING THE LOW GAME

FIRST games always contain a number of challenges for teams to tackle. In 2006, the challenges included gathering balls from the playing field, using the camera to locate the vision target above the goal, shooting balls into the upper goal, depositing balls in the lower goal, and climbing the ramp at the end of the game.

The annual goal of Team 322 is to pick one aspect of each year's game and be the very best at that aspect of the competition. By focusing on one game challenge, full attention can be devoted to that part of the game and a simpler, more robust robot can be designed. Such an approach requires the wisdom to select the right design feature around which to build the robot, and the discipline to stick with that decision. The intense nature of the short FIRST season makes such discipline all the more important.

Team 322 is a large and diverse group. The team consists of students and teachers from Flint (Michigan) Central, Flint Northern, and Flint Northwestern-Edison High Schools, and mentors from the University of Michigan-Flint and General Motors. This team typifies the education and employment pipeline that FIRST promotes to its participants: from high school to college to engineering corporations.

## ⇄ Design Philosophy: Do One Thing Better than Anyone Else

Team 322 identified the following goal for the 2006 FIRST Robotics Competition: to be the most mobile, reliable, and efficient ball-herding robot in the game. The team would not attempt to design a robot for the more difficult task of shooting balls into the upper-goal, but instead it would be the best low-goal scoring robot. By being the best at that task, the team believed it could provide a strategic advantage to any alliance.

Picking one item to focus on is not as easy as one might think. One of the team's hardest tasks was keeping team members convinced they had chosen the best design. The team believed it could be at least three times more efficient at low-goal scoring than the best upper-goal scoring robot. Since each ball scored in the lower goal earned one point and each ball scored in the upper-goal was worth three points, Team 322's approach was sound. But part of Team 322's strategy called for them to be the best lower-goal scoring robot; quite a challenge, given that over 1,100 robots participated in the 2006 FIRST Robotics Competition.

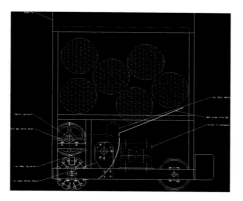

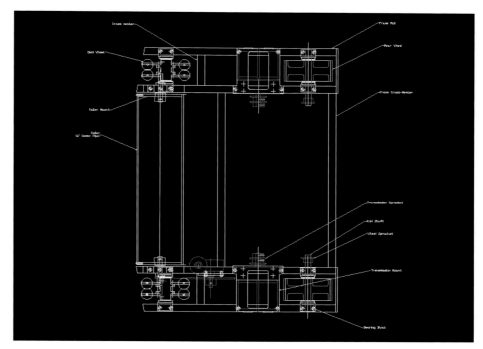

⇄ ↓↓ Designed with a single purpose, to score in the lower goal, the robot includes a large hopper to carry balls. To maximize the robot's ability to gather balls, the robot is constructed with its largest dimension across the front of the robot.

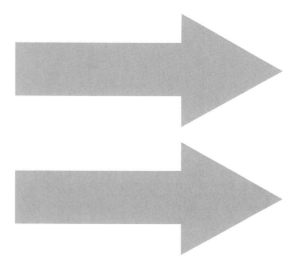

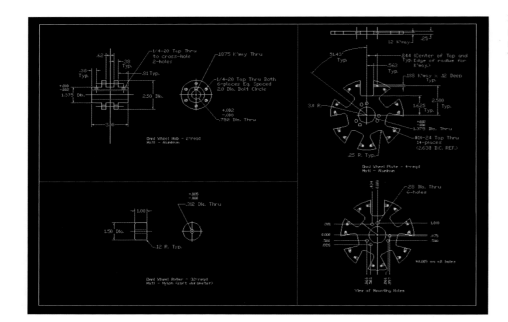

## ⇄ Promoting Thinking over Tinkering

In addition to their clear focus on a single attribute, Team 322 was successful because of their approach to planning. The team used a strict schedule to guide their work during the six weeks of the FIRST Robotics Competition build season in January and February. The first few days were devoted to understanding the game itself, selecting a strategy of play, and coming to agreement on the final design decision. For Team 322, this period was very important; if the team decided on the wrong strategy the robot would not be competitive.

The next week was spent drawing the robot using CAD tools. This part of the design process documented all aspects of the entire robot, including each subsystem (such as the drive train) and part (such as the wheels) that needed to be manufactured.

The frame and robot drive train were the first components designed since the rest of the robot was built on this foundation. Construction of the base began the following week. Also during this third week, the remaining components were modeled in the computer. At the end of the third week, the detailed robot design was finished, and a complete set of design drawings for the entire robot was created.

The robot was constructed the following week, with the fifth and sixth weeks of the design period used for driving practice, electronics, programming, and debugging the robot and its program. This final period of the build cycle was critical to the team's success as it provided opportunities to test and improve the robot's performance and robustness. During this period, programmers tested different autonomous modes, and the robot was challenged with as many obstacles as possible to identify weak aspects of the strategy or design.

↑↑ Diamond plating protects all robot systems, including the wheels and internal electronics. The low profile design, with the majority of the weight concentrated in the robot base, is very stable.

# ⇌ A Low-Goal Powerhouse

Team 322's robot was designed to be a ball harvester and low-goal scorer. Without a superstructure, the design was compact, protected, and low to the ground. Beneath the outer diamond-plate armor were solid systems for propulsion, ball gathering, and delivery. Notable components that benefited from the team's prior competition experience included the drive train, traction wheels, omni-wheels, and ball harvester.

Team 322's robot was a powerful, four-wheel-drive machine. Two motors were used with each transmission provided in the Kit of Parts to power the front and rear wheels. One chain drove the robot's traction wheels and a second chain powered the omni-wheels. The traction wheels produced high levels of torque and the omni-wheels increased the robot's maneuverability.

The only alteration made to the stock transmissions was reducing the weight of some of the internal gears by machining away excess material. The team was careful not to overdo it when reducing material to make a part lighter. In a prior year's competition, Team 322 learned the hard way that taking away too much material can lead to structural failure of the component. In 2004, the team reduced a significant amount of material on a sprocket to save weight, only to have that sprocket fail during a competition because of a lack of structural strength. As such, caution was used with the 2006 weight reduction measures.

The traction wheels were custom-made by Team 322 and machined from a solid block of aluminum. The wheels were designed for performance and to minimize the number of parts used for assembly. By reducing the number of parts, the team increased the reliability of the wheels. The wheel rims were covered with commercial conveyor-belt material for high traction.

Team 322 designed the omni-wheels used on their 2006 robot. Popular with many FIRST teams, omni-wheels use embedded rollers to reduce the frictional force during turns. The rollers are mounted perpendicular to the wheel axis and allow the wheel to roll sideways during turns. Prior omni-wheels constructed by Team 322 used polyvinyl-chloride (PVC) pipe as the rollers. The spacing between the PVC rollers resulted in substantial bouncing as the wheel rotated, thereby producing a constant shock load on every robot part. Also, the slippery PVC surface did not provide sufficient traction.

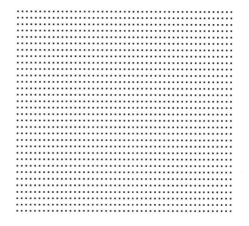

⬇⬇ Dual sprockets on the transmission output shaft allow separate chains for the forward and rear drive wheels, introducing redundancy and increasing the system's reliability.

⇒ Caution is used when trimming parts to reduce weight. In the 2004 FIRST Robotics Competition, one of Team 322's sprockets failed because of the loss of strength that resulted when the sprocket was machined to reduce its weight.

↑↑ The rear traction wheels are machined from solid blocks of aluminum. The wheels use a minimum number of parts and mounting fittings to reduce the possibility of component failure.

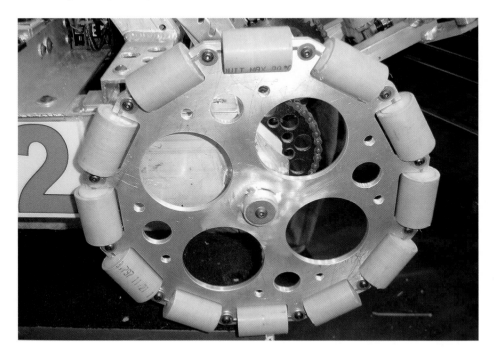

↑↑ FIRST Team 322's original omni-wheel design used PVC rollers distributed around an aluminum hub. The PVC did not provide sufficient traction.

To improve the prior design, the PVC rollers were replaced with solid urethane rollers and each composite wheel was constructed with a pair of wheels. Following the simplicity principle used with the traction wheels, the omniwheels were designed for easy assembly and high reliability. The pairing of wheels produced a smoother ride and avoided the bouncing associated with the single wheel design. The urethane rollers were offset from each other in the installed wheels to produce a nearly contiguous urethane rolling surface.

The ball-retrieval capabilities of the robot garnered significant attention, given the team's focus on scoring in the lower goal. The robot was designed with the largest dimension along the front to provide the widest possible opening for gathering balls and depositing them in the lower goal. A single roller served as the ball harvester, and the collected balls were stored in the large ball hopper at the top of the robot.

As with the drive wheels, the team manufactured hubs for the ball harvester, rather than modifying a less-than-optimal commercial component. A sprocket-and-chain power train was used to rotate the roller at a high rate of speed. To drop balls in the lower goal, the roller rotated in reverse. This system was simple and fast: important design attributes for a machine that was designed to be the best in its class.

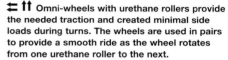 Omni-wheels with urethane rollers provide the needed traction and created minimal side loads during turns. The wheels are used in pairs to provide a smooth ride as the wheel rotates from one urethane roller to the next.

⬆⬆ ⇄ The wide, large-diameter roller extends across the face of the robot. The roller continually spins forward to ingest balls from the field, and rotates in reverse to shoot balls into the lower goal.

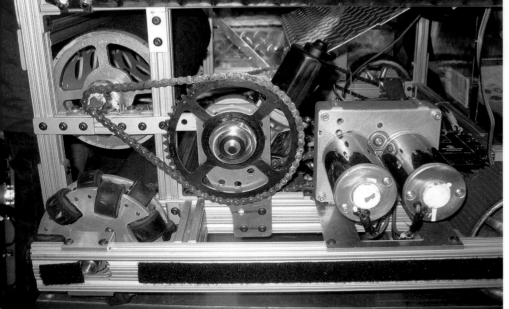

⇄ Gearing on the ball-pickup system increases the rotational speed of the collecting roller by a factor of six. High rotational speed increases the roller's ability to gather balls.

# ➨ Robustness Ensures Success

Three factors from Team 322's design methodology produced a robot that withstood the competition. One was the aluminum extrusion frame that formed the robot base. Assembled with nuts and bolts, the frame was less rigid and more forgiving than a welded frame. That lack of rigidity became an asset whenever the robot was hit during the competition as it allowed for slight movement and avoided cracking at points where members joined together. Repairs usually consisted of straightening parts and tightening loose fasteners.

Adopting successful design attributes from prior designs was the second factor contributing to robustness. The transmissions from the Kit of Parts were selected because the team had found them to be very durable during the previous year's competition. Prior experience also reinforced the need to maintain a low center of gravity. By mounting major components low in the robot, the robot's stability improved and it could reliably climb the ramp without toppling over.

Guarding the robot's components was the third factor that contributed to its robustness. All of the components were shielded to protect them from damage by other robots. Diamond plate guards surrounded the robot shell to prevent damage to internal robot components. The shielding was easy to remove to access the internal robot mechanisms.

Team 322's ability to identify one aspect of the game at which to be the very best paid off handsomely. The team was a formidable competitor. The robot was ruggedly built, not prone to damage, and easy to maintain. The drive-system was fast and highly maneuverable, and the ball gathering/scoring system was exceptionally effective. This combination of simplicity and ruggedness resulted in a robot that was one of the very best, just as the team had planned from the very start.

➨ Structural aluminum forms the robot frame, and a diamond plate protects the robot from damage. The diamond plate also adds to the robot weight, improving traction.

# QUALITY BUILT: TREADS, TUBE, AND TURRET

## DESIGN INNOVATIONS FROM TOP TO BOTTOM

Team 1024 wisely applied its three years of robot-building experience to their entry in the 2006 FIRST Robotics Competition. Over that time, the team realized that to win matches a robot had to withstand a harsh competitive environment and remain fully functional throughout the competition. Past experience also taught the team that a competitive robot needed to be equally effective on offense and defense.

The creative process relies on openness and advances with fresh ideas. Team 1024, from Indianapolis, Indiana, generated new ideas with a brainstorming session focused on robot qualities and attributes. Small groups discussed the game and developed concepts that defined three qualities for a competitive robot. Group consensus confirmed that a winning robot had to be strong, sturdy, and have an accurate shooter.

Subteams were formed to examine robot performance related to mobility, functionality, and control. The team decided that the robot needed to be dominant on defense and quick and accurate while playing offense. In terms of game features, the team wanted a robot that could score in the upper goal, climb the ramp at the end of the game, and hold its own on the field.

Based on the list of desirable robot qualities and functions, the team created a list of design features for the robot. The drive system would include a two-speed transmission powering tank-style treads. The shooting system attributes included an ability to shoot ten balls in eight seconds, an aiming system independent of the robot base, and the capability to regulate the firing velocity and angle.

The list of qualities and attributes led to design innovations from the top to the bottom of the robot. The robot would include creative design features in three dimensions: a tread-driven propulsion system, tubular construction, and a top-mounted turret. With a unique propulsion system in the base and a superstructure made of high-strength tubes capped with a rotating turret at the top, the entire robot showcased novel design. Collectively these components satisfied the attributes, qualities, and goals established in the initial brainstorming phase of the design process and resulted in a competitive machine.

## ⇉Design Innovation 1: Powerful Propulsion and Treads

The drive system was the first design innovation. Years of experience benefited the team's development of an effective propulsion system. The requirements to be quick on offense and strong on defense mandated that the propulsion system be designed for multiple game conditions. To accomplish these two goals, the propulsion system included four powerful motors provided in the Kit of Parts and a two-speed transmission.

In low speed, robot torque was maximized to give the machine power to push other robots out of scoring position and easily climb the ramp. In high speed, the robot could quickly traverse the field to gather balls and maneuver into scoring position. The treads were selected because of their advantages in pushing power and climbing ability.

An open-bay frame, needed to collect balls from the field, dictated the propulsion system arrangement at the rear of the robot. Identifying such a requirement at the beginning of the planning process later benefited the team with its goal for a low center of gravity to improve the robot's stability when climbing the ramp.

⇆ Concepts are evaluated and design constraints examined with foam and cardboard mock-ups of a robot design. These materials are easy to work with and are used to create full-scale models of preliminary concepts.

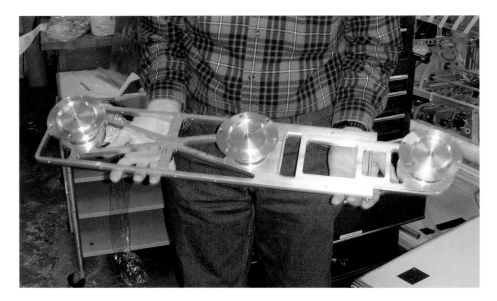

⇄ Each side of the drive train features rollers mounted on a sliding rail. The rail slides to provide tension on the drive belts and to adjust for stretching as the belts wear.

Refinements over three years were manifested in the track drive system for the 2006 competition. The track system used a single rail on each side of the robot to support the track's pulleys and bearings. A slide, similar to a trombone slide, allowed the track to be easily tensioned. The side-mounted rails defined an entry area for gathering balls, and offered the needed space at the rear of the robot to mount the propulsion hardware. The rails also served as the foundation for the rest of the robot.

Large-diameter, slotted plastic wheels were mounted to the rails, with the dual-speed transmission powering only the rear wheel. A flexible timing belt connected the three wheels. Because the timing belt had protrusions on both sides it could transfer power from the four propulsion motors to the playing surface without slipping. This propulsion system met the team's goal for high speed and significant pushing power.

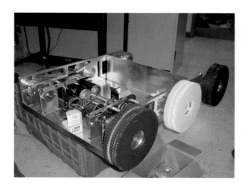

⇈ An advantage of a tread design is that only one wheel in the drive train needs to be powered. The remaining wheels are free-spinning and rotate as the tread passes over them.

⇈ Treads provide great traction and can climb over obstructions. With ridges on both sides, the treads grip both the wheels and playing surface.

# ⇉ Design Innovation 2: Tubular Support

The second design innovation materialized in the robot's midsection. The design challenge was to create a structure that connected the shooting mechanism at the top of the robot to the robot base. This structure needed to be strong to withstand collisions with other robots and to provide a solid foundation for other robot systems.

Motivated by the need to keep the center of gravity low, the team was inspired by a similar structure used for protection on go-carts and dune buggies. The idea of a roll-cage for the robot was suggested. Such a structure could be formed with lightweight aluminum tubes and would provide a strong platform that would protect the robot and be suitable to mount other components on.

Tubing is a sound, but seldom used, material for robot construction. The structure created by Team 1024 was constructed with 1-inch (2.5 cm)-diameter aluminum tube. An aluminum plate was welded to the base of the tubes, and this plate was mounted on the robot base. The open design of the tubular frame met the design goal to keep the center of the robot open for a ball gathering and lifting system.

The most important feature of this structure was its strength-to-weight ratio. Weighing in at just more than 4 pounds (1.8 kg), the fame provided the necessary shooter height without making the robot top heavy. Lightweight cables at the front and rear of the frame and small aluminum plates along the sides provided rigidity to the structure. Because each frame piece was bent from a single piece of tube, combined with the strength of the circular cross-sectional shape, the resulting structure was strong and solid.

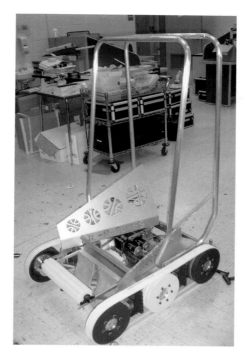

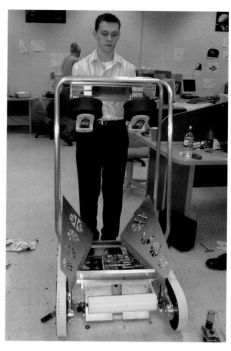

↑↑ A roll cage, fabricated with aluminum tube, provides a rigid frame to protect the robot. Lower side panels help direct balls into the elevator and increase the cage's rigidity.

↑↑ The shooting system is suspended from the roll cage. The open nature of the roll cage allows other auxiliary systems to be added to the design.

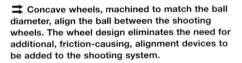

➡ Concave wheels, machined to match the ball diameter, align the ball between the shooting wheels. The wheel design eliminates the need for additional, friction-causing, alignment devices to be added to the shooting system.

## ⇉ Design Innovation 3: Suspended Turret

The third design innovation was the shooting system located at the top of the robot. Differing from most designs that had the turret resting and rotating on a lower surface, Team 1024 designed their turret to be suspended from the tubular frame above it. For this configuration, the ball delivery system was fixed and the shooting wheels rotated around a central delivery point. The design made the turret and loading systems independent and enabled a nearly continuous feed of balls to the shooter. Because of the minimum amount of mass of the suspended system the shooter could be quickly aimed.

↑↑ A single bearing supports the suspended shooter from the overhead frame.

⇄ The shooter is suspended from above and freely rotates. The design allows the shooter to be fed from a fixed, central, position, and does not require the feeding system to rotate.

A series of prototypes led to the final design. The initial prototype was powered with standard wheels and used a cardboard tube to position the balls between the shooting wheels. This system demonstrated an advantage of using two wheels to propel the ball and illustrated the need to carefully align the balls in the shooting system.

A refinement to this prototype replaced the original wheels with a set of concave wheels designed and constructed by the team. The surface of the new wheels was machined to match the ball diameter. The shape was chosen to align the balls and equally distribute the load as each ball passed through the wheels. These factors improved the accuracy of the shooting system.

Although the initial plans for the shooting system called for an ability to adjust the aim in both horizontal and vertical dimensions, tests indicated that 90 percent of all shots could be made with a fixed, horizontal angle. This design decision greatly simplified the shooting mechanism and resulted in a system that could hang from the overhead structure and be aimed with a single motor.

A bearing plate was designed to support the suspended turret. Two high-precision bearings were sandwiched between aluminum plates to reduce friction when the turret rotated. The turret drive sprocket and attached shooting system were suspended from the bearings.

Target location was detected with a camera mounted on top of the shooter and the turret angle was measured with a potentiometer attached to the turret. Closed-loop feedback control minimized the difference between the two signals and compensated for any disturbances caused by collisions during shooting.

## ⇄ Great Design from Top to Bottom

The suspended shooter, aluminum support frame, and tread drive system formed a robot that met the team's desires for strength, endurance, and accuracy. Team 1024 benefited from their well-planned design process which included multiple prototypes, computer modeling, and rigorous system testing.

The team regularly tested individual components while they were being developed, rather than waiting for a complete robot assembly to be finished before testing. As a result of working along parallel paths an effective machine was developed, more people were engaged in the design process, and the best robot possible was created in a short period of time.

⬆⬆ ⬇⬇ Balls collect in the lower hopper before they are lifted in the rear of the robot to the shooting mechanism. Mounting the camera and the shooter at the highest points on the robot reduces the chances for the mechanisms to be blocked by opponents.

# SECTION 05

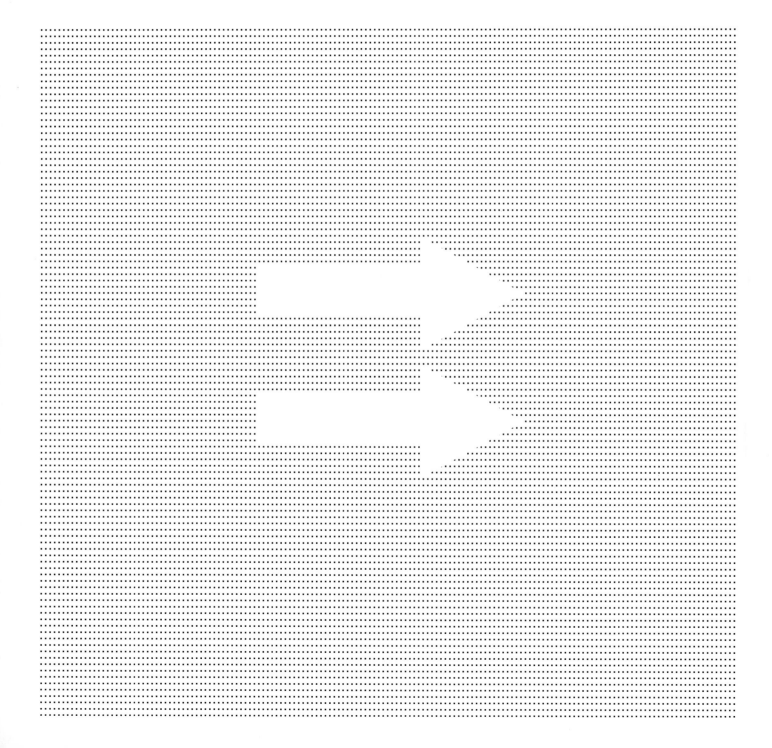

The Xerox Creativity Award celebrates creativity in design, use of a component, or strategy of play. While one team may be recognized for a single creative component, another team may have won this award for its creative method of scoring points. The award promotes ingenuity and originality. The award showcases unique designs that offer a competitive advantage. Teams that win this award progress from a creative idea to a creative product, and that journey is often not an easy one.

Xerox is a founding sponsor of FIRST. The company provides generous team and regional event support as part of its long-standing commitment to education and as an investment in the future diversity of its workforce. A former chairman and CEO of Xerox is also chairman emeritus of the FIRST board of directors, and the current senior vice president and president of business group operations now serves on FIRST's Board.

# Xerox Creativity Award

# A NEW APPROACH TO A FAMILIAR DESIGN

## ENDING THE OPEN BIN'S BAD REPUTATION

Many teams focus on a robot's ability to store game balls. In 2006, the idea of an open storage bin was often abandoned because the foam balls easily stuck to each other and any containing walls, causing frequent jamming. The random arrangement of the balls was also unfavorable, especially if they needed to be delivered one at a time to a scoring mechanism. Team 86, from Jacksonville, Florida found a creative way to work around the limitations of an open-bin design.

## ⇄ Building a Robot in a Garage

Team 86 students lead the design, fabrication, and programming of each robot. The students train year-round in all of the skills needed to build a winning robot—including the use of Autodesk Inventor software, machine tools, and welding equipment. The team also works year-round on its programming ability.

Team 86 had amongst its sponsors two different types of engineering companies. Vistakon, a division of Johnson & Johnson Vision Care, Inc., is a contact lens manufacturer; JEA is the municipal electric, water, and wastewater utility for Jacksonville. Although one might assume that Team 86 conducted its magic at one of its sponsors' advanced machine shops, the work was done closer to home, in a mentor's home workshop. The team was structured so that each student developed at least one specialty skill in welding, lathe or mill operations, computer-aided design, website development, or computer programming.

## ⇄ Strategy Starts With the Snorfler

Team 86's 2006 robot was designed to quickly collect balls from the playing field floor. This was accomplished with a rotating collector located at the front of the robot. The collector was constructed from PVC pipe and operated at a speed of 3,200 rotations per minute. The collector was almost the entire width of the robot, and this enabled the device to collect two side-by-side balls at once. Acknowledging the ability to rapidly retrieve balls, the collector was named the Snorfler.

The balls collected from the floor were transported to the top of the robot by a series of polyurethane cords, driven and guided along PVC rollers. Once the balls reached the top of the robot, they were deposited into a central bin. This is where the creativity of the students really came into play.

⬇⬇ **PVC rollers were attached to an aluminum frame. These guided the polyurethane cords which transported the foam balls to the top of the robot.**

⇈ A bin prototype demonstrats how easily the foam balls would stick to each other and to the bin walls.

⇄ A stirring mechanism is incorporated into the bottom of the bin prototype, which consists of cords wrapped around rotating PVC pipes and set at an angle.

⇈

## ⇄ Origin of the Bin Design

During the initial brainstorming phase, the team members agreed they wanted a robot that could quickly score goals. In order to facilitate this plan, they needed a mechanism to store a large number of balls. They chose a bin-style storage system as they found that linear storage systems did not have the desired ball capacity or delivery speed.

While most of the team were debating a way to optimize the sidewall angles of the bin to prevent jamming, one student suggested incorporating a stirring mechanism to prevent the balls from sticking to the sides of the bin. Initially, other team members were skeptical because of the stirring mechanism's complexity and additional cost, weight, and power drain on the battery. To convince the team of the feasibility of such a design, the student constructed a working prototype. The prototype worked well, and on a final design sketches began.

Four aluminum wedge-shaped walls enclosed the storage bin. The ball collector and polyurethane cord system, which carried eight balls per second up and into the bin, was located outside the bin's front wall. Outside the rear wall of the bin, another polyurethane cord system carried balls up to a high-goal shooter.

The bin's floor was key because it kept the balls from sticking to each other and to the bin walls. The floor was constructed from a set of rollers and polyurethane cord placed at an angle. When the rollers were set in motion, the balls would roll around and avoid jamming against one another and the bin walls. The bin capacity was 32 balls. These balls could be fed rapidly to score in either the upper or lower goals.

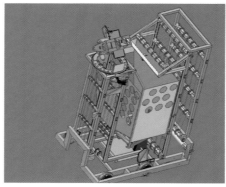

⇄ ↓↓ A drawing and a photo of the robot depict the polyurethane cord systems that delivered the balls from the ball collector into the bin and from the bin into the shooter.

↓↓ A front view of the robot shows the slanted walls of the bin located behind the ball collector and transport system.

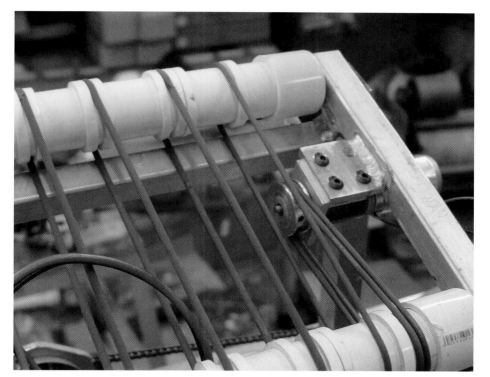

⇄ ⇊ Seen here are a close-up view of the rotating floor component with the motor that drives it, and a view looking through the side of the bin at the floor rollers.

⇈ A team member fits a foam ball against the conveyor system which feeds balls out of the bin to be scored in a goal.

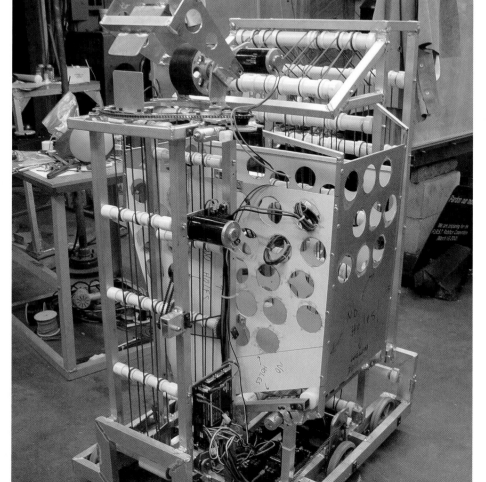

⇄ This angled view of the robot provides a glimpse of the floor rollers positioned at the bottom right of the bin.

# ⇉On to Scoring

When it came time to deposit the balls into a goal, a conveyor belt guided them to another roller that transported the balls. If this roller rotated in one direction, the balls would be shot out from low on the robot into the lower playing field goals at a rate of five balls per second. When the roller direction was reversed, the balls were sent up the rear conveyor system to a turret-mounted camera-guided shooting mechanism. The shooting mechanism could fire the balls into the high goal at a rate of three balls per second.

⇊ A conveyor belt guides the balls from the bin to a roller system that would further transport the balls.

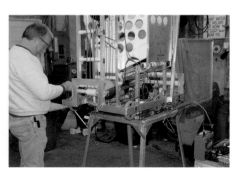
⇊ Once the balls pass through the conveyor belt, a system of rollers and cords send them either out the bottom of the robot or up to the shooting mechanism.

⇄⇊ An outlet at the back of the robot permitted balls to be scored in the lower goals, depending on the direction of the roller spin.

## ⇄ No-stick Bin Design

The bin was not only physically but also metaphorically the central part of the robot. Its ability to store a large quantity of balls was key to the success of the ball collection and delivery systems. The innovative method for stirring the balls to prevent sticking and jamming allowed the other mechanisms on the robot to perform smoothly.

Although only a small team, the students on Team 86 have consistently produced the most creative ideas. They followed through on their creative solutions and designed, fabricated, assembled, programmed, and tested inventive mechanisms.

Team 86's solution to the ball-storage problem resulted from a student's imagination. By designing a way to overcome a problem faced by many teams, Team Resistance lived up to its name and team motto: "Going against the current." All in all, admirable results were produced a small, home-based workshop.

↓↓ Despite its seemingly complex design, Team 86's robot utilizes a unique and effective way to transfer balls.

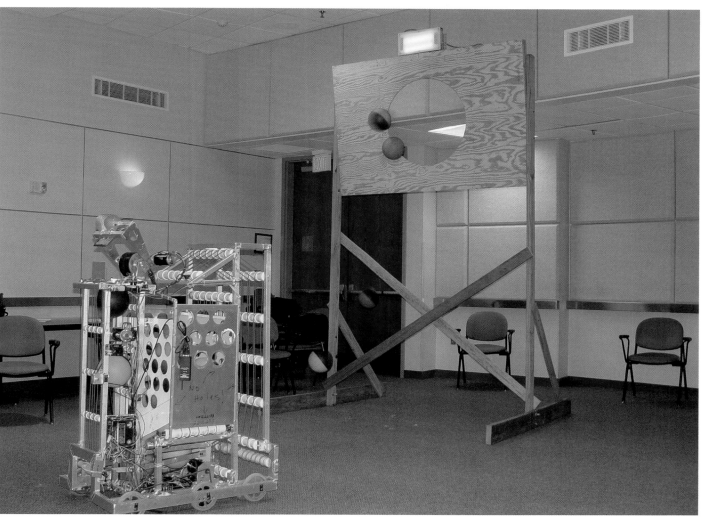

# EXPANDING POSSIBILITIES

## HELPING THE HUMAN PLAYER

Some teams start the design process by focusing on one feature and concentrate their efforts to maximize performance of the chosen attribute. Team 293, The Bullbots, from Pennington, New Jersey, studied the game rules and discovered that one of the rules could be applied to give the team a tremendous advantage. The creative design of this robot produced a characteristic feature, making Team 293's robot easy to recognize on the field and distinguishing it from other competitors.

The team decided to use the human player to load the robot with balls that would be shot into the high goal. To load the robot, the human player had to toss balls over a 7-foot (2.1 m) partition into the robot. It was not expected that the robot would remain stationary during loading; other teams would constantly interfere to disrupt the robot's ability to receive potential scoring opportunities. Given the potential difficulty loading balls, Team 293 focused its attention on making the loading target as large as possible.

# ⇉ An Expanding Net to Capture Shots

In the 2006 FIRST Robotics Competition game manual concerning robot size limitations, there was an article which stated: "Once a match begins, robots may extend horizontally beyond the 28-inch (0.71 m) x 38-inch (0.97 m) starting size under their own power, up to a limit of 60 inches (1.5 m) in either horizontal cardinal dimension. The robot may not exceed the 60-inch height restriction at any time during the match. Any restraints (elastic bands, springs, etc.) that are used to maintain starting size must remain attached to the robot for the duration of the match." Team 293 decided to capitalize on this rule to make it easier to load balls, a decision that influenced the entire robot design and contributed to their success in winning the Xerox Creativity Award.

Since the robot could expand once a match started, the team reasoned that its robot should expand in a way that maximized the target size for the human player to aim at. A net was designed that could be folded around the robot at the start of each match, unfold on command, and expand to the maximum size allowed by the rules. The team anticipated this expansion occurring during the autonomous mode of the competition, thereby making the robot available for loading at the beginning of the tele-operated play period.

⇊ A net on the robot expands at the start of each match, providing the human player with a much larger target than the original footprint allowed.

# ⇌ Building the Net

To construct the net, the team created a design and a pattern was drawn on paper. The net top was designed with a subtle arc that provided 4 inches (10.2 cm) of sag over the net's 5-foot (1.5 m) length. Four panels were created; one for each side of the net. A tube was fastened to the upper perimeter of each panel to accommodate a cable that supported the net. This cable ensured that the net would not expand beyond the 60-inch by 60-inch (1.5 x 1.5 m) extension size limit.

The arc-shaped upper edges of the panels prevented sagging and helped uniformly distribute the load on the support cable. The final net used on the robot was constructed from black, vinyl-coated fiberglass screen, a material that was lightweight, flexible, and resistant to tearing. An additional attribute of the material was that the screen material did not restrict the view of the robot operators. Black Lycra wings attached the seams of the net to the support poles. These wings helped stabilize the poles and centered the net. To finalize the design, the seams and edging were reinforced with cotton cloth.

The net was supported by a frame made with lightweight aluminum rods mounted on springs attached to the robot frame. The springs played a vital role in the design, as they prevented penalties against the robot if the net collided with other robots. The game rules only allowed forcible robot-to-robot contact within the robot's starting volume. The collapsible spring design protected the robot from being penalized each time the expanded structure made contact with another robot. The springs also ensured that the net was redeployed when the applied forces were removed.

Surgical tubing was added to the springs to apply 30 pounds (133 N) of tension, enough force to ensure that the net would stay open, or reopen if it were pushed. The springs allowed rod movement in all directions, thus improving the system's robustness and reducing its vulnerability to damage. The rods were mounted to the robot frame to create the largest possible net opening, and to position the top of the net at the maximum allowed height. In addition to its utility for scoring, the net also worked as a defensive tactic to block shots from other robots while being loaded with balls by the human player.

⬇⬇ Four panels of net were sewn together and lined at the top with cable to maintain the dimensions of the fabric. Black wings at the bottom attached the net to support poles.

⬇⬇ The support poles are drawn to calculate size and the required spring angle to maximize the flexibility of the system.

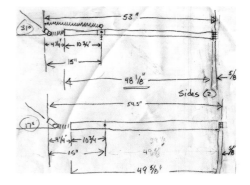

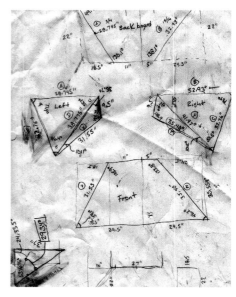

⇌ ⬆⬆ Before construction of the net began, preliminary sketches were made to determine the exact dimensions and shape of the system.

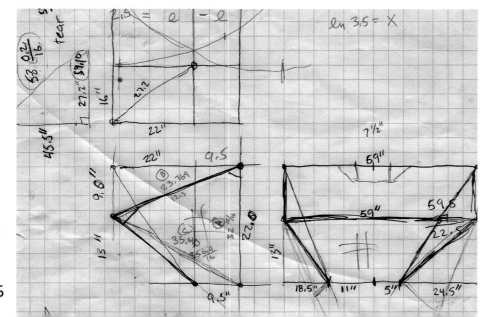

# ⇄ From the Net to the Goal

With the net design complete, the details for transferring the stored balls to a shooting system remained to be determined. The team decided the shooter should be located as high as possible to avoid defensive interference from opponents. As such, a mechanism was needed to transport balls from the bottom of the net to the shooting system. Team members considered many different ways to accomplish this, but each design they came up with did not completely meet their needs, was impractical, or was too complicated. When they had almost given up hope of designing a feasible system, inspiration suddenly hit.

Earlier, the team had discarded an idea for using a single-wheel shooting mechanism and instead elected to proceed with a two-wheel model. But just because the performance of the single-wheel device was not adequate for shooting did not mean it couldn't be used in another application. The team realized that a single wheel, with guides placed along a perimeter, could effectively transfer the foam balls simply by rotating.

Prototypes were made of this transport system, and the design evolved into the form that was incorporated on the final robot. Two single wheels, vertically stacked and backed by S-shaped guide rails, proved to be an effective and reliable transport method to move balls from the bottom of the net to the shooter. This shape provided the height to transfer balls to the shooter, maximized the use of space in the robot, and contained enough room to store balls before a match. If the balls jammed in the mechanism, the wheel rotation could simply be reversed to free the balls from the system.

A smooth transition from the net to the beginning of the S-shaped shooter feed system was needed to prevent the balls from jamming at the entrance to the transport mechanism. To create this transition, a foam mold was created and Acrylonitrile Butadiene Styrene (ABS) plastic was cast around the mold. The resulting lightweight plastic basin averted ball jams and enabled a smooth flow of balls into the shooter feed system.

Two infrared light gates were mounted along the balls' path from the net to the shooter. One sensor signaled the operator that a ball had entered the feed system, and the other signaled that a ball had arrived at the top of the feed system and was ready to shoot. This monitoring system assisted with controlling the number and position of balls in the robot.

A dual-motor, four-wheel shooting system topped the robot. The motors were directly connected to the shooting wheels which rotated at a high rate of speed. By using two motors for shooting balls, the system could quickly recover momentum after each shot, and had the capability for rapid scoring.

⤓⤓ A single-wheel prototype proves that balls could be effectively transferred, provided they contacted an opposite plate, seen here as the plywood frame.

The evolution of the ball-feed system is shown: from a rough sketch, to a more precise drawing, to incorporating the feed device and shooting wheel in a CAD drawing, to a working prototype of the full system.

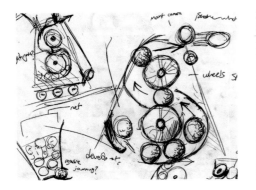

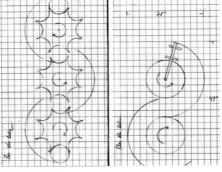

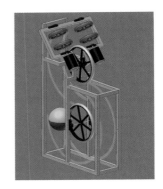

A side view of the unfinished robot clearly shows the black, molded-plastic basin on the right which enabled a smooth ball transition from the net to the feed system.

# ⇄ Making a Bigger Target

Team 293 designed the robot frame to support the innovative net and shooting system. The entire robot structure was first fabricated with insulating foam to visualize the size and layout. Structural aluminum was used as the construction material to build the base and robot frame. The use of a common material simplified the overall design.

Knowing that their team's performance would rely heavily on the success of the human player to load balls, Team 293 maximized the ball target on the robot. By taking a close look at the game rules the team members were able to take advantage of their creativity and build a device that met the rules and satisfied their design goal to ease loading. The team's unique game strategy, in which the robot simultaneously loaded balls while preventing other teams from scoring, resulted in a notable performance on the field.

⇊ The Bullbots created a unique robot design by taking a closer look at the rules of the game.

# CREATIVE DESIGN FROM THOROUGH PLANNING

## ROBOTIC CHUTES AND LADDERS

Following a regimented design, construction, and testing process is a necessary requirement to produce a robot in six weeks for the FIRST Robotics Competition. One team in particular developed a methodical and exceptionally thorough approach to the design process. By mimicking a design process used in industry, the team was able to promote creativity, planning, and teamwork while building the robot.

Team 330, from Hope Chapel Academy in Hermosa Beach, California, is a team that thrives on imagination. Its entry in the 2006 FIRST Robotics Competition was especially creative, with an integrated system of conveyors, lifts, and ramps that in one aspect resembled the childrens game Chutes and Ladders. Each ball moved through 14 vertical feet (4.2 m) as it passed through the lifting, storage, and transport systems from entry to exit. The robot stood out for its well-integrated system and for the compact packaging that united the independent systems.

## ⇄ Following an Industrial Design Process

Team 330 approached the FIRST challenge using a systems-engineering design process used throughout the engineering industry. The process is a systematic method to recognize the initial design requirements and generate ideas through brainstorming and strategizing. To gain consensus, the team reviewed all ideas before agreeing on a final design.

Upon learning the specifics of the "Aim High" challenge, team members dove into the design process by first reading and understanding the game rules. To better understand the game, a full-sized model of the playing field was marked out, and the students played out different possible strategies. A replica of the playing field was drawn on a large table and the team members simulated different games using cardboard robot models.

These activities allowed team members to study all aspects of the competition before thinking about specific machines that could play the game. By observing how the matches could play out, team members were able to establish sound game-play strategies and to create valid initial design proposals.

## ⇄ Advancing to a Design

Preliminary and critical design reviews ensued, where conceptual drawings were presented and discussed. Once the final design was selected by a vote of the entire team, prototyping and computer modeling commenced. Autodesk Inventor was used to create drawings of the robot to visualize volume allotment and possible component interferences.

Strategies and designs continued to be examined and tested until an idea solution was agreed upon. The system of continual design review and improvement resulted in a well thought-out initial design. By using this systems engineering design process, all robot details were closely reviewed and finalized before any fabrication took place. This system prevented mistakes and preserved resources. More effort was put into brainstorming and prototyping, because the team believed that the best solution is rarely discovered on the first try, but rather results from a continual improvement process.

The central feature of the robot was its ball conveyance mechanism. Balls were pulled from the playing field floor into the frame of the robot through two 1-inch (2.5 cm) -diameter PVC rollers rotating in opposite directions. When a ball made contact with these rollers, the ball was compressed between them and pulled into the robot.

The motor that rotated these rollers also ran a fabric conveyor belt equipped with small rubber teeth that provided sufficient grip on each ball.

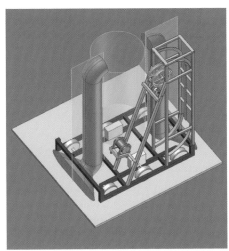

⇈ A rough Autodesk Inventor drawing laid out the robot components so the team could get a better idea of the space allotment.

⇄ A tabletop replica of the playing field was drawn out, and cardboard robots enacted different possible strategies.

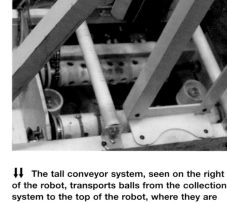

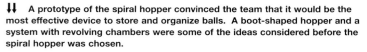

⇄ The two horizontal, PVC rollers can be seen here. Their rotation caused balls to be pulled into the robot's frame.

⬇⬇ The tall conveyor system, seen on the right of the robot, transports balls from the collection system to the top of the robot, where they are guided by the curved frame into the hopper.

⬇⬇ A prototype of the spiral hopper convinced the team that it would be the most effective device to store and organize balls. A boot-shaped hopper and a system with revolving chambers were some of the ideas considered before the spiral hopper was chosen.

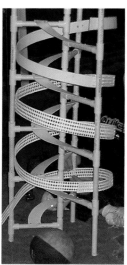

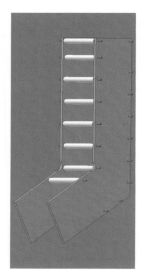

The conveyor belt pulled the balls up the five-foot high aluminum frame and dropped them into the top of a four-foot high, helix-shaped storage hopper.

The spiral hopper was chosen as the preferred system from a collection of other proposed hopper designs, including an upright boot-shaped hopper, a large open bin, a bin with an incorporated ramp system, and a revolving chamber mechanism. Each design was prototyped and compared to determine the most effective storage system.

The helix design was favored because it kept the balls ordered in a linear fashion, and had a large holding capacity of 20 balls. A PVC skeleton frame bordered by Lexan walls supported the helical ramp and contained the balls within the storage system.

## ⇄ Advancing to Either the Shooting Wheels or Lower Goal

The balls delivered to the top of the hopper by the conveyor were gravity-fed to the bottom of the spiral. From there, the balls were conveyed one more time using belts and pulleys back to the top of the robot where the shooting wheel was located. To keep the weight distribution low, the shooting-wheel motor was mounted near the robot base, and a belt was used to transmit power to the shooting wheels at the top of the robot. A camera guided the shooting mechanism and aligned the robot with the goal.

If the direction of the conveyor belts was reversed, the balls would roll along another conveyor belt to the lower front end of the robot to be deposited in the lower goals. Thus, either high or low goals could be scored using the same mechanism. This feature was an important design criteria identified at the beginning of the design process.

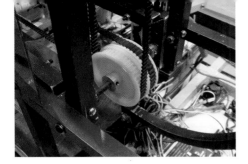

⇈ Belts and pulleys assist in transporting the balls from the bottom of the spiral hopper to be scored in either the high or low goals.

⇈ A camera-guided shooting mechanism, made of two wheels rotating on the same shaft, assist in locating the high goal.

⇊ A conveyor belt guides balls from the bottom of the spiral hopper and out the front of the robot to be scored in the low goals. The outlet for these balls is seen to the right, above the ball collection system

# ⇄ Sound Design Leads to Sound Construction

The optimum velocities of the rollers and belts were calculated. Gears and pulleys were selected to drive each system at a specific rate: the balls traveled from the PVC rollers to the top of the hopper in approximately 1 second, down the spiral hopper in 2 seconds, and up to the shooting mechanism in 1.5 seconds. This fast travel rate meant efficient processing of the balls and maximized Team 330's capability to score points.

The systems engineering design process applied by Team 330 began by uniting all team members to achieve results. Although a small team of only ten high school students, Team 330 achieved much, earning the Xerox Creativity Award at two separate regional competitions. The team believed that creativity came from exploring all possibilities and keeping an open mind to all ideas. By systematically planning and prototyping, Team 330 was able to produce a unique robot that incorporated an intricate and effective ball transport system.

⇄ The intricate design of the storage system, with all of its belts and pulleys, effectively performs its desired task.

# BRISTLE-POWERED BALL TRANSPORT

## A DIVERSE TEAM CREATES INNOVATIVE SOLUTIONS

Many times, challenges arise during the construction of a FIRST robot that are unforeseen in the planning phase. The ability to work around the unexpected can mean the difference between failure and success on the field. Team 694 from Stuyvesant High School in New York City designed their robot with a creative method of conveying balls through each mechanism, and when they faced a difficult challenge they were able to think of alternatives and implement a design solution.

Innovarive ways of thinking stemmed from the team itself and its diverse group sponsors, which included Credit Suisse, D. E. Shaw and Co., Yvette and Larry Gralla, Con Edison, Time Inc., The Wallace Foundation, Cox and Company, the Stuyvesant High School Alumni Association, and the Parents Association. The diverse group of team members stimulated creative thinking.

↑↑ Four sets of rotating hairbrush bristles pull balls into the front of the robot, and a clear ramp direct them into the hopper.

## ⇉ Moving Foam Balls from The Floor to The Robot

The plan for the 2006 robot incorporated ways to collect, store, and score foam balls. The front frame of the robot would be left open to incorporate a ball collector to pull balls from the playing field into the machine. The balls would then be diverted to a storage device and delivered to a single-wheel shooting mechanism located at the top of the robot.

An early idea to collect the balls off the playing field floor suggested that the bristles from a hairbrush could provide the required grip on the foam balls.

This idea was investigated by placing hairbrush bristles on a motor-driven rotating drum. The tangential velocity of the rotating bristles was equivalent to the robot's maximum-drive velocity, and the combined speeds advanced balls into the robot.

Once collected by the bristles, a quarter-circle-shaped ramp directed the balls into the spiral hopper. When the direction of the rotating brushes was reversed, the balls could be fed out the bottom and into the lower goals. The bristles proved to be an effective means to move the balls in and out of the robot.

⇊ A spiral prototype was constructed out of
pool noodles supported by outer vertical
aluminum channels.

⇈ A paper prototype of the spiral exhibited both
strength and lightness, as the ramp and outer
wall reinforced each other.

## ⇉ You Can't Score if You Don't Store

Several designs were considered for ball storage. The team decided a rolling conveyor belt would not have the desired storage capacity and a square bin could result in jamming. They discovered that a spiral hopper would provide a large amount of surface area on its coiled path to store balls, and it would not be prone to jamming as the balls progressed along the coil in single file.

The spiral bin was first prototyped to determine its feasibility. The original, open design consisted of vertical shafts protruding from the circumference of the base of the spiral. Netting and plastic sheeting enclosed the system and kept the balls from falling out. To conserve weight, the design was modified and other outer skeletal materials explored, such as pool noodles and rubber tubing. A paper and styrene model was constructed to test the strength needed to support multiple game balls. The testing revealed that by combining two thin materials to reinforce each other, a strong and lightweight structure resulted. These results were applied to create the final design for the ball storage hopper.

⇊ With the final hopper design chosen, a **CAD** model is drawn to better visualize what the final product should look like.

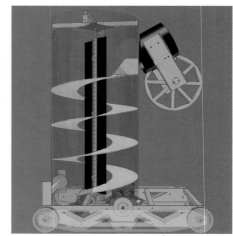

A helix was constructed out of light-weight-aluminum donut-shapes. These shapes were slotted, stretched into the helix, and riveted together to form the extended helix. To convert the flat aluminum cut-outs into the correct size, shape, and constant diameter to complete the helix, the spiral was viewed as a series of slanted lines that intersected an outer shell. Using the Pythagorean theorem applied to the height between sections and the distance from the center to the outer edge of the helix, the correct donut diameters were calculated.

Sheet metal rings with 90-degree tabs were attached to the outer edges of the helix. These tabs were riveted to, and supported, a Lexan shell, and both the helix and shell were mounted to the chassis. The hopper was created out of Lexan because it was a strong and lightweight transparent material.

The design of the hopper and spiral formed a complete assembly, and no part could be removed without disassembling the entire system. To provide easy access into the hopper, especially if the robot had no, circular holes were cut into the shell, one at each level of the ramp. These holes were large enough to remove balls if needed but were normally covered with fiberglass disks. These disks were painted with the team number and attached to the helix shell with temporary fasteners.

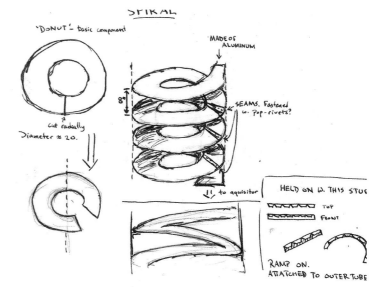

⇈ This sketch shows the process of creating the helix, from the donut-shaped aluminum pieces to the final spiral.

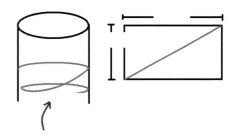

⇄ The Pythagorean theorem helps relate the spiral shape to a rectangular relationship between the ramp circumferences and the height between each ramp level.

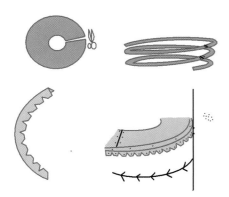

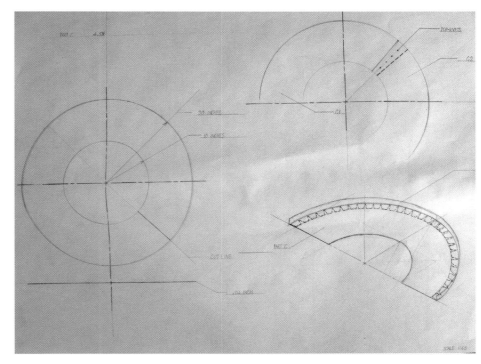

⇋ ⇈ Affixed to the outer rim of the helix ramps were sheet metal rings with tabs fixed at a 90-degree angle. These tabs were riveted to a Lexan shell that encased the helix.

# OUTER TUBE

### CONSTRUCTION

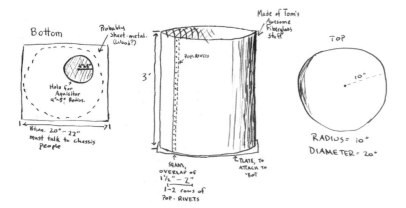

**Bottom**
Probably Sheet-metal. (Wood?)

Hole for Aquisiter 4"-5" Radius.

Btum. 20" - 22"
must talk to chassis people

Made of Tom's Awesome Fibreglass Stuff

POP-RIVETS

3'

SEAM, OVERLAP of 1½" - 2"
1-2 rows of POP-RIVETS

PLATE, To ATTACH To 'BoT

**TOP**
10"
RADIUS = 10"
DIAMETER = 20"

⬇⬇ The clear Lexan shell provids a view of the ramps inside the hopper. The driver can easily determine if any balls are present inside.

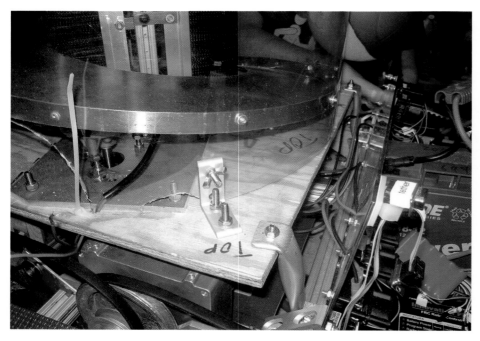

⬆⬆ Along with being riveted to the ramps, the outer shell is also attached to the robot chassis with L-shaped brackets.

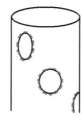 

➡ Holes placed along the sides of the hopper provide easy access to the inside of the helix. The Velcro disks that cover these holes are attached well enough to keep the balls inside the hopper, but can also be removed easily for inner access.

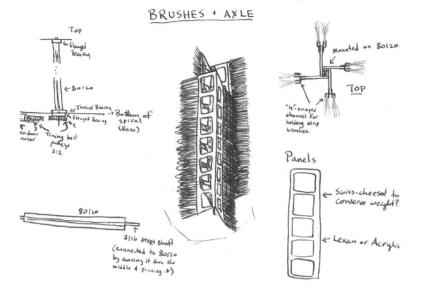

BRUSHES + AXLE

## ⇄ Getting from Bottom to Top

To impart motion and move the balls up the spiral, the brush idea used in the front roller was applied once again. Long bristles were attached to a rotating shaft located in the center of the helix. When the shaft rotated, the balls were pushed up the helix and into the shooting mechanism, where they were ready for scoring.

When the balls reached the shooting mechanism, they would stop climbing because of the presence of a small plastic slope at the entrance to the shooting mechanism. This assisted in pausing the balls at the top of the mechanism while still allowing balls collected at the bottom to move up the spiral. If the brushes ran in reverse, the balls were fed down the spiral and out the collector to score in the lower goals.

The shaft was centered in the hopper with top-mounted aluminum crossbars and directly mounted to the chassis at its base, where it was powered by a motor and belt/pulley system. The length of the bristles and rotational speed of the shaft were optimized for the most efficient ball movement.

↑↑ Brush bristles are attached to a rotating shaft centered inside the helix. The rotation of this shaft pushes and guides the balls up the spiral.

⇐ The plastic slope can be seen to the right of the shooting wheel. This prevents balls from traveling further up the helix.

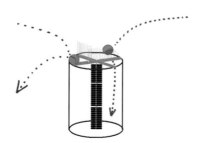

The spiral hopper had a capacity of 28 foam balls. These balls could be loaded from the playing field floor or be loaded into the top of the robot by the human player. Testing revealed that when the human player threw balls into the open top of the hopper the balls tended to bounce out, especially when they hit the aluminum crossbars that held the brush shaft in place. To correct for this, holes were drilled into the crossbars and green plastic spokes were individually glued into these holes.

The resulting effect looked like grass growing out of the top of the robot. These spokes helped cushion the ball and caused it to land softly in the hopper, rather than bouncing out. Occasionally, the balls shot by the human player hit the rim of the hopper shell and again bounced out. Thin aluminum bars were attached to the top of the hopper and bent outwards. A spider-web system of twine and knots connected these bars and helped guide the balls into the hopper.

The unique implementation of the green plastic spokes and spider web proved to be effective. Although they may not have looked like common components for a robot, they successfully fulfilled their purpose.

⬆⬆ ⇉ The green fibers at the top of the robot assist in cushioning balls that are thrown by the human player, preventing them from bouncing out. The spider web that prevented balls from bouncing off the rim of the hopper is also visible at the top of the robot in the photograph.

## ⇄ Bristles to Advance and Grass to Contain Ball Movement

Team 694 benefited from creative ideas that were analyzed and improved. During their evolution from idea to product, concepts were optimized for performance. The use of bristles to convey balls came from one simple suggestion, and this single thought ended up being the main identifiable feature of the robot. Team 694's success illustrates that creative ideas can come from anywhere, and that having an open mind to possibilities encourages the development of innovative solutions.

⇐ Team 694 used creative materials to overcome design obstacles they faced while building their robot. These contributed to the unique look of the machine.

# DOUBLE REDUNDANCY AND TRIPLE-HOPPER OPTIONS

## EARLY PLANNING ACCELERATES CONSTRUCTION PERIOD

Attention to detail in the early phases of robot development can prove to be very beneficial as the end of the six-week construction period nears. Team 1319, from Mauldin, South Carolina, designed and built a robot that exhibited creativity in both its drive train and its comprehensive ball delivery and scoring mechanism. In addition to being innovative, the systems were also designed with redundancy to improve their on-field performance.

To ensure a continuously working robot, the drive was designed to remain functioning in the event of a hardware failure. Instead of designing and building one mechanism to transfer balls, three systems were built that each completed this task in a different way. With extensive planning and prototyping at the beginning of the build period, the construction process was accelerated and allowed for additional time to develop these multiple design solutions.

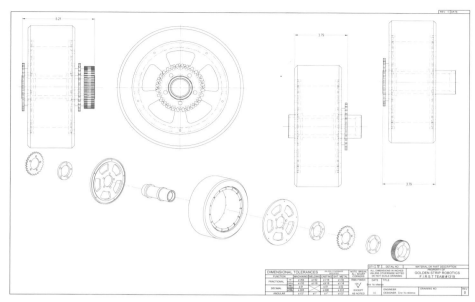

↑↑ A drawing of one of the custom-made wheels shows how it is put together with the gears and chain-driving sprockets.

⇐ The team name was etched into the custom-lasered wheels.

⇄ A close-up of the drive train shows the gears, sprocket, and chain used to connect the directly-driven center wheel with the outer wheels.

## ⇄ A Doubly Redundant Drive System

The initial desired features for the drive train included reliability, speed, and strength. Accomplishing all of these features in one design, in a limited time, could be considered an intimidating task. However, Team 1319 found a way to include all of these features by designing a double-redundancy drive system.

This drive system featured three wheels on each side of the robot. A single motor, coupled with a three-speed transmission, drove each of the center wheels. Servo motors were used to shift the transmissions.

The center wheel on each side powered each of the outer wheels with a separate chain. If one of the chains were to break during a match only the outer wheels connected by it would be disabled, and the robot would still be able to maneuver. If all chains broke, the robot would still be mobile because the center wheels were directly connected to the motors. Having planned for multiple failures while preserving functionality, the propulsion system

came to be called the double-redundancy drive.

The use of separate chains for each wheel anticipated and corrected for possible component failures. This was especially important for the "Aim High" game because of the probability of aggressive play.

The six wheels were custom made with a width of 3 inches (7.6 cm) to yield high amounts of traction. With multiple high-traction wheels, the drive system had the power to climb the ramp and the ability to maneuver with ease on the playing field.

# ⇉ A Winning Scoring System

Because climbing the ramp was deemed an important part of winning the game, planning had to ensure that the robot would not tip over. The center of gravity needed to be as low as possible. It was known that a shooting mechanism with its own separate drive motors would contribute a significant proportion of the robot's overall weight. The decision was made that such a mechanism would have to be placed low on the robot, and that the ball hopper, being lightweight, would need to be placed higher.

A unique design approach was used for the main scoring mechanism. To be adaptable to different match conditions, a device was needed that could score in both the low and high goals. A shooting mechanism capable of firing high and low was constructed to meet this design goal. The shooter was mounted on a turret ring that rotated 270-degrees. The turret ring was custom made, belt driven, and rode inside grooved bearings.

The turret position was tracked by a potentiometer, and the measured signal was relayed to the robot controller. To protect the wires that provided power to the turret–mounted mechanisms, a flexible, lightweight hose was used to encase the wires and keep them from tangling in other robot components. This hose expanded and contracted as the turret rotated.

A shooting wheel, driven by a pulley-and-belt system, accelerated and launched the foam balls. A device referred to as the "clam shell" adjusted the exit angle of the balls to score in either the low or high goals. Slots were cut into the "clam shell" to adjust the distance, and, as a result, the compression between the shell and the shooter wheel. The vertical tilt angle of this shell was adjusted by a gearbox mechanism powered by a small motor. A potentiometer tracked the tilt angle.

⇊ The turret and shooting wheel are seen upside down here, providing a clear view of the turret ring and belt-driven pulley at the top.

⇈ The turret ring rode inside grooved bearings to allow for easy rotation throughout its 270-degree range.

➡ The lightweight hose used to encase the wires that ran up to the mechanisms on the turret was very flexible.

➡ The "clam-shell" component cradled the shooting wheel. Its position could be adjusted to change the launch angle of the foam balls.

## ⇄ Three-Hopper Mechanism for Three Modes of Game Play

The ball-storage container, fabricated from lightweight Lexan and aluminum angle, was located above the shooter wheel and "clam shell," and rotated with the turret. The team designed a three-hopper mechanism to be used on the robot.

In one design, a sophisticated mechanism in the hopper advanced balls to the shooter wheel. A two-level hopper, capable of being loaded by the human player, was divided into five chambers by a paddle. This paddle was motor-driven to rotate within the hopper. A ball-sized hole, located in the bottom of the hopper, led to the "clam shell" and shooter wheel. With each revolution of the paddle, one ball would drop into the shooter. A photo switch was positioned over each paddle to monitor its position and ensure that the paddle stopped in the correct position for a ball to drop into the hole.

A second hopper mechanism was also divided into five chambers and loaded by the human player, but it could feed two balls at a time into the "clam shell" and shooter wheel.

The third hopper design consisted of a conveyor system with a ceiling. Balls traveled up and over a conveyor and were dropped into the shooting mechanism. Unfortunately, this idea did not provide a sufficient ball storage capacity and could not be loaded by the human player.

During the regional competition, each hopper design was used, and all worked well. The most effective one turned out to be the second design, which fed two balls at a time into the "clam shell." Based on its superior performance, this design was chosen to be implemented on the robot at the FIRST Robotics Championship.

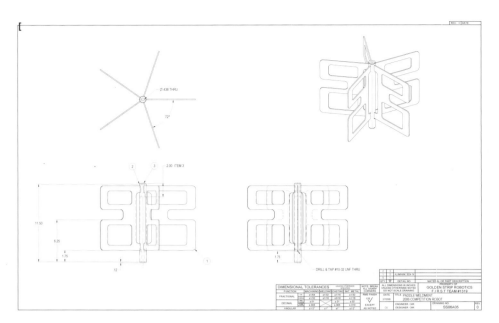

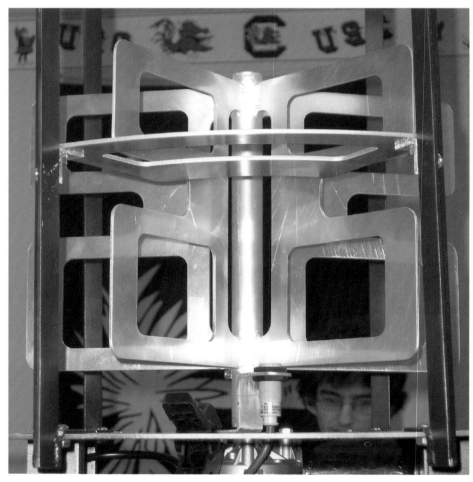

⇈ A paddle divided the hopper into five chambers, each of which could hold a ball. This paddle rotated with the hopper and fed a ball into the shooting mechanism with each rotation.

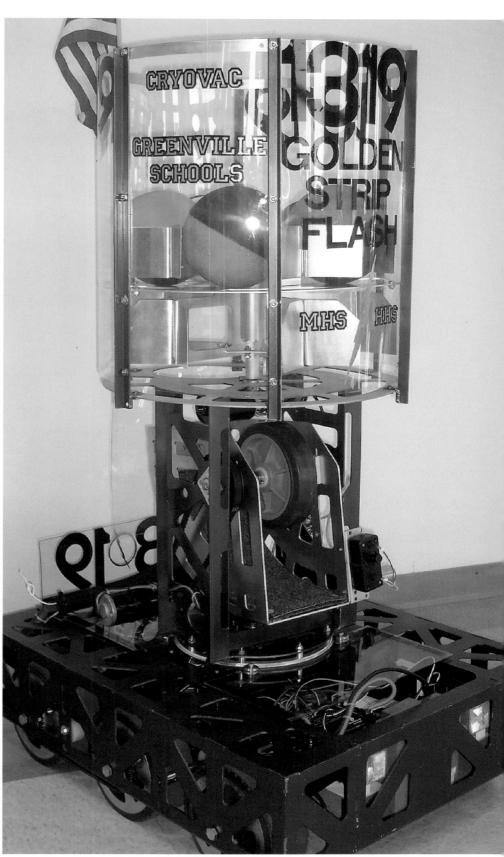

Team 1319 built a versatile robot with a modular hopper that could be interchanged with different designs.

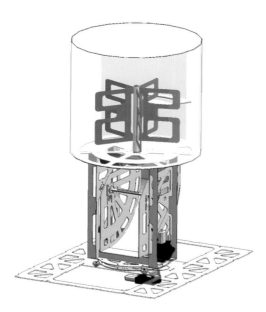

A drawing shows the relation of the paddle to the clam shell and shooting mechanism. Balls that exited the hopper were gravity fed to the shooter.

The methodical process of brainstorming, prototyping, and precise, three-dimensional modeling allowed the team to identify and correct possible design flaws early in the build process. The lessons learned from prototyping and modeling eased the transition from concept to reality. The resulting robot used a fault-tolerant drive system and could incorporate three alternative designs to manipulate balls within the robot. The team's innovative and creative abilities produced redundant systems that increased reliability and multiplied the team's chance for success.

# EVOLUTION AND ADAPTABILITY

## MODIFYING A DESIGN TO REACH PERFECTION

The final version of a FIRST robot does not always resemble the original plans. In fact, it is actually rare that the final robot exactly resembles the original design. Changes must be accommodated as more is learned about physical, material, feasibility, and programming restraints. A lot of new information is gathered throughout the build process, and designs need to adapt to modifications and changes.

Design changes require the team's constant attention. Every member can contribute creative ideas that improve the design. With the motto "Student Designed, Student Built, Mentor Approved," Team 1510 from Beaverton, Oregon, stated their preference for student participation in the process.
The hands-on experience of building a robot requires that the students persist through the entire process, including all changes and modifications that led to the end product.

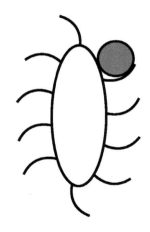

## ⇄ Initial Designs and Initial Discoveries

Team 1510's original design for the robot included standard features found on many robots: a chassis and drive train for mobility, a ball gatherer to collect balls off the playing field floor, a turret-mounted shooting mechanism to score in the high goal, a hopper to store balls, and an electronics and control system to integrate each device. Once each component had been defined and planned, prototyping commenced, and three-dimensional computer models were created.

The first two mechanisms designed and constructed were the shooting and gathering systems, deemed most important for the robot's success. When the final parts of these systems were dimensioned, the team realized its original plans for the ball-storage system would have to be modified, as the space for such a component was very limited. A new system would need to

take up less space, as well as be simpler and lightweight.

Several ideas were originally considered to collect the balls off the playing field floor. Some of the suggested ideas were a ridged conveyor belt, a front ball catcher using a wedge ramp, and a dual conveyor belt to squeeze the balls from each side. To store the balls, external baskets, cloth bags, and vertical tubes were proposed.

When they discovered there would be limited room to incorporate a full-hopper structure, the team had to return to the drawing board. A mechanism was needed to unite the ball gatherer and the shooter. The physical restraints of the systems left only 9 inches (22.9 cm) between the ball gatherer and the ball shooter. The new collector would have to include the turret-mounted shooter, be able to load balls into the shooter, regardless of the position of the turret, and not cause the balls to jam.

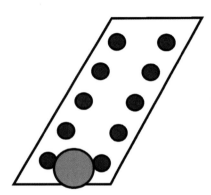

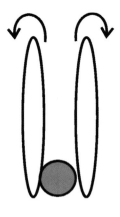

⇈ Some designs presented for the collection of balls were a cranberry conveyor belt, a front ball catcher with a wedge ramp, and a front ball gatherer.

## ⇄ Designing a New Solution

The final composite ball-manipulation device resulted from a simple re-design that incorporated all of the newfound constraints. The resulting ball gatherer served the dual purpose of gathering and storing balls. The gatherer was built to extend nearly the entire allowed width of the robot, permitting maximum room for ball storage. A front roller was designed and machined from PVC pipe that pulled balls from the floor onto a small flexible ramp.

⇄ A wide, front gatherer maximized the number of balls that could be collected off the playing field floor.

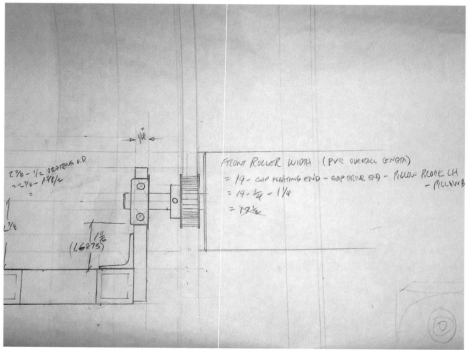

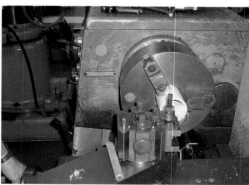

⇄ ⇈ A sketch of the front roller was used to machine the device out of PVC pipe. The rotation of this roller pulled balls into the robot with the help of a small ramp.

transported

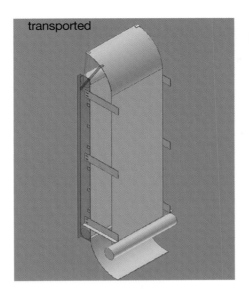

⇈ A drawing of the wide conveyor belt that transported balls to the top of the robot shows the front roller and ramp at the bottom.

⇉ The transported balls were pulled up the robot and were sandwiched between the moving conveyor belt and a clear polycarbonate sheet.

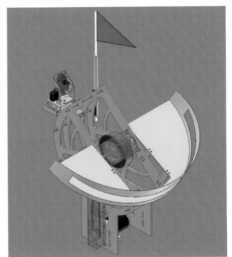

⇈ After exiting the conveyor belt, the balls were guided by a curved fender into a semi-circular bin which channeled them into the shooting mechanism.

Once inside the robot, the balls were transported by a wide but lightweight conveyor belt. The balls, sandwiched between the conveyor and a polycarbonate sheet, rolled along the polycarbonate sheet inside the robot. At the top of the robot, balls dropped out of the conveyor and were guided by a curved fender into a semicircular structure attached to the shooting mechanism. This shape collected balls dropped at any turret angle, allowing the shooter to be loaded regardless of turret position.

The conveyor system was designed to run in two directions: forward mode and reverse mode. The forward direction guided balls up to the shooter for scoring in the upper goal. In the reverse direction, balls flowed out of robot and into the lower goal. An optical sensor controlled the starting and stopping of the conveyor belt, incrementally advancing the conveyor only when a ball was detected entering the gatherer. This allowed for the stacking of up to 12 balls in the conveyor system.

⬆⬆ The shooting mechanism was mounted atop a camera-guided turret. The bin allowed balls to be loaded to the shooter regardless of its position.

# ⇄ Adapting Designs to Meet Rules

Team 1510 viewed the constraints and rules set by FIRST not as a hindrance to creative thinking but rather as an instigator for promoting innovative thought. Competition rules require teams to think creatively to find solutions that work within a given set of guidelines. Weight limitations require investigation into new materials, and size restrictions force creative solutions to accommodate many mechanisms. The six-week time period to complete all tasks demands that creative solutions be focused on the task at hand. The constraints compel each team to find unique, creative solutions to the design challenge and result in a wide variety of solutions to the same problem.

The evolution of the ball gatherer illustrates the creative ability of Team 1510 to adapt to unexpected changes. To create a comprehensive system that met the FIRST rules, the ball-gathering device went through many modifications from its original design. To meet the physical size restraints, a single mechanism was designed to perform the duties previously handled by three mechanisms. The ball-gathering device implemented on the robot gathered, stored, and delivered balls to separate locations. The team's flexibility and their ability to overcome the unforeseen resulted in a design that was simpler, yet as effective, as the original plan.

Guided by the motto "student designed, student built, mentor approved," Team 1510 operated with a structure that they feel best promotes student learning. Team 1510's success, as demonstrated by the students' ability to understand and solve problems, demonstrated the validity of their motto as a tool to motivate students and deepen their understanding of design.

⇈ The size restraints encountered by the team actually aided in simplifying the final robot design. Team 1510 was able to learn and adapt throughout the build process.

# AFTERWORD

## by Woodie Flowers

⇆ ↓↓ Professor Woodie Flowers is a cofounder of the FIRST Robotics Competition and an international authority on engineering design. While providing autographs to FIRST participants, Flowers collects the signatures of FIRST students and mentors on his own shirt. This action is a reflection of his admiration for their work with FIRST.

# FIRST:
# MUCH MORE THAN ROBOTS

This book showcases how good design results in impressive machines. These robots are the product of creative collaboration between students and mentors. The designs are the result of a process that taps into unrealized talents of students and their mentors, bringing them a new confidence that is often life-changing.

The machines are impressive, but it is the effects of the building process that keeps me working as a volunteer for FIRST. With the Kit of Parts and a goal in place, the people in a FIRST group first have to become a team. Watching that team form, learn, and produce is immensely rewarding. A variety of talents come to the fore to solve problems. Students with technical skills work with those with an artistic talent, and the members with organizational abilities tie it all together. Students find in themselves depths of ability that they did not know they possessed. The mentors guide and support the students and, in turn, get the warm feeling that comes with helping others and the pride that results from your team's accomplishments.

Working long and hard under the pressure necessary to complete the projects, we learn to communicate and to cooperate. We learn that, properly done, teamwork produces professional products and personal satisfaction. Working in a FIRST team is a microcosm of solving much larger and more daunting problems in the world.

FIRST supports a concept of "gracious professionalism." Gracious people respect others and let that respect show in their actions. Professionals possess special knowledge and are trusted by society to use that knowledge responsibly. Thus, gracious professionals, acting with integrity and sensitivity, make a valued contribution to society.

"Gracious professionalism" describes the FIRST spirit of encouraging high quality, well-informed work in a manner that leaves everyone feeling valued. "Gracious professionalism" seems to be a good descriptor for part of the ethos of FIRST. It is part of what makes FIRST special and wonderful.

I believe that people like a tough challenge. In fact, FIRST is founded on that belief. FIRST offers young people a unique opportunity to be challenged, to think like professionals, and to solve complex engineering problems. At the same time, they learn personal skills they will call upon throughout life. Teamwork, creativity, and sociability—all skills that help adults succeed in making positive contributions to society—are essential parts of the FIRST experience.

As an MIT professor, I have had opportunities to get involved in many challenging problems. FIRST is one of my most consuming challenges. It is also one of the most rewarding.

*Dr. Woodie Flowers is the Pappalardo Professor of Mechanical Engineering at the Massachusetts Institute of Technology, a Distinguished Partner at Olin College, and cofounder of FIRST's cornerstone program, the FIRST Robotics Competition. Professor Flowers participates in the design of the FIRST Robotics Competition game each year and has served as a National Advisor to the FIRST Robotics Competition since its inception.*

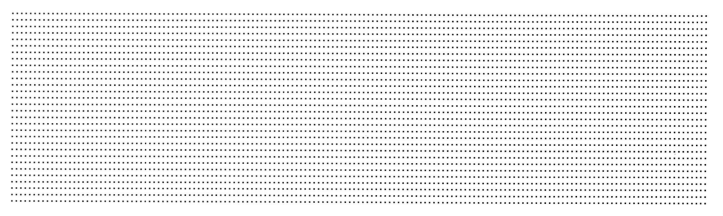

# ⇄ Directory of Award Winners

## Delphi "Driving Tomorrow's Technology" Award

### Team 25: Bristol-Myers Squibb and North Brunswick Township High School

**North Brunswick, NJ**

Wayne Cokeley, Bob Goldman, Michael Lubniewski, Charles Smaltino, Walter Suchowiecki, Kevin Durham, John Dusko, Anthony Kukulski, Shaun McNulty, Michael Schroeder, Mario Azar, Michael Palazzo, John Chester, Bruce Ciance, Bharat Nain, Corey Balint, Alison Rankin, Srirag Babu, Matthew Bauer, Matthew Beckles, Shoham Bhadra, Kristian Calhoun, Lisa Chen, Keith Chester, Michael Ciance, Tina Ciance, Anthony Ferrer, Morgan Gillespie, Kirsten Guevarra, Samantha Hires, Yiling Hu, Shane Ogunnaike, Aditya Pandyaram, Neil Parikh, Dev Patel, Hasit Patel, Hannah Petersen, Vishal Ramani, Mike Savin, James Schroeder, Jigar Shah, Christopher Signoretti, Robert Slattery, Alexa Stott, Vijay Venkataraman, Maria Xu, Mike Zankel, Bryant Lam, Jesse Linder, Chris McLean, Brian Mollica

### Team 33: DaimlerChrysler and Notre Dame Preparatory

**Auburn Hills, MI**

Tim Grogan, Jim Zondag, Pamela Williamson, Tricia Stauber, Isaac Rife, Jason Monroe, Chester Fleming, Eric Yahrmatter, Matt Spicuzzi, Mike Kaurich, Kathryn Siuniak, Amanda Siuniak, Andrew Woodcox, Mike Gower, Erik McDonald, David Huston, Brian Fain, Carla Spicuzzi, Jennifer Doll, Sarah Yahrmatter, Christina Winkler, Sean Grogan, David Wulbrecht, Scott Karbon, Keith McDonald, Connie Pulliam, Kevin Thompson, Phillip VonDonop, Jerry Yahrmatter, Carol Belkowski, Larry Jack

### Team 100: PDI Dreamworks, A Better Mousetrap Machining, Woodside High School, and Carlmont High School

**Woodside, CA**

Janet Creech, Arlene Kolber, Laura Robeck, Peter Adams, Gerardo Aguilar, Emma Cambron, Rachel Cape, Brad Carrender, Billy Chartain, Christian Constantz, Geoff Dubrow, Tevye Friedlander, Steven Gordon, Luke Gray, Johnathan Greer, Kevin Irish, Ben Kahn, Riley Kelley, Jonathan Kurniawan, Alex Morgan, Max Mudd, Mike Pinto, Michael Rhodes, Steven Rhodes, Nathan Roleff, Stephanie Ruane, Kiro Salama, Jeshua Sosa, Trevor Sturm, Martin Taylor, Roman Tetelman, Andrew Theiss, Chris Wuydts, Danny Aden, Richard Burgess, Jim Cape, Eva Carrender, David Frydenlund, Bob Kahn, Cory McBride, Laura Rhodes, Christine Taylor, Al Theiss

### Team 357: PECO Exelon, ITT Technical Institute, and Upper Darby High School

**Drexel Hill, PA**

Megan Durkin, Joe Troy, Michael Crane, Beth Hale, Jules Scogna, Adam Schuman, Chris Troy, Andrew Adaman, Lauren Campion, Melissa Cell, Alex Crane, Sreee Dandamudi, Kevin Durkin, Jeffrey Davis, Joe Adams, Christina Ghilardi, Joe Gro, Sid Joshi, Patrick Kneass, Jon Knippschild, Bob Knowlton, Jen Knowlton, Maddi LaSage, Chris Massi, Rachel McKlindon, Nikki Paltit, Ben McCarron, Taif Choudhury, Jason Cola, Roger Singh, Garret Sapsis, Dan Troy

### Team 467: Intel and Shrewsbury High School

**Shrewsbury, MA**

Elizabeth Alexander, Brittany Alphonse, Vanessa Alphonse, Devin Anderson, Andrew Arena, David Arena, Sarah Arena, Mitchel Baker, Mark Belanger, Michael Castelli, Anna Chan, Carolyn Cooper, Rebecca Cooper, Jeffery Davario, Darren Del Dotto, Brendan Donoghue, Nora Donoghue, John Easson, Muneeb Elsakka, Laura Friedman, Julianne Jensen, Daniel Jones, Eric Jones, Sean Keegan, Benjamin L'Bassi, Anthony Ly, Sathyajith Maliakal, Kate O'Brien, Christina Oullette, Ankur Patel, Caitlin Russell, John Slavkovsky, Edward Smongeski, Alan Tran, Timofey Trubko, Steve Ward, Tom Ward, Matt Whalen, Joshua Freier, Erich Dieffenbach, Raj Patel, John Tsombakos, Robert Cornacchioli, Richard Cooper, Dottie Jensen, Jim Jensen, Cathy Cooper, Nancy Donoghue

### Team 1501: UTEC, Wabash Technologies, 4H Robotics, and Huntington North High School

**Huntington, IN**

Seth Boyd, Rob Brumm, Jacob Christman, Matt Crosley, James DeHaven, Branden Gage, Steve Hite, Kevyn Hollowell, Larry Johnston, Brittany Jones, Sam Kratzer, Nosh Nave, John D. Paff, Ryan Ploughe, Kassi Reppen, Krystal Reppen, Chris Riddle, Chris Routzahn, Austin Shaw, Cameron Sivley, Steven Wilcox, Chris Brumm, Tim DeHaven, Wayne Doenges, Casey Drudge, Chris Elston, Curt Henderson, Harley Henline, Sonny Henline, David Kratzer, David Kuiper-Moore, Jerry Smyth, Mike Smyth, Al Thorn, Scott Thorn

## General Motors Industrial Design Award

### Team 121: NAVSEA Undersea Warfare Center, Raytheon, University of Rhode Island, Middletown High School, Portsmouth High School, Tiverton High School, and Mount Hope High School

**Newport County, RI**

Preston Anderson, Bridgette Blight, Jamie Bova, Christian Sullivan, Andrew Cochrane, Michael Fenton, Liana Ferreira Fenton, Tony Formica, Joshua Gabriel, Mitchell Hathaway, Patrick Marks, Timothy Marois, Tyler Messinger, Alex Monteiro, Will Neederman, Jim Oconnor, Dave Ferreira, Ryan Miller, Evan Patton, Evan Poitras, Christopher Rinier, Matt Shea, Jon Sirr, Joe Teno, Florian Uhbac, Stacie Waleyko, William Walker, Phillip Weston, Jeff Whalley, Rick Casey, Rob Zeuge, Tom Dolan, Tom Barron, Curt Oldford, Dave Nugent, Cadence Ellington, Mike Fenton, Liana Ferreira Fenton, Gary Gabriel, John Patton, Tom Frank, Brad Allspach, Chris Jackson, Geoff Sitnik, Joe Menassa, Kyle Fenton, Matt Shea, Mike De Sousa, Rick Blight

### Team 123: Ford, General Motors, Coffey Machining Services, ITT Tech, and Hamtramck High School

**Hamtramck, MI**

Joe Julian, Joanne Malczewski, John Stofflett, Amy Zacharias, Muamer Abdurahmanovic, Jasenka Bektas, Kaleigh Borushko, Dewan Choudhury, Megan Cook, Catherine Gasior, Syed Didar Hussain, Mohammed Jalal, Zakaria Juber, Mariola Koplejewski, Hana Mehmedovic, Indira Memisevic, Justin Ostaszewski, Shakil Qureshi, Shaw Sabith Qureshi, Joshua Saganski, Sharmin Salam, Monika Toton, Admir Trnjanin, Ashley Tulecki

### Team 190: WPI and Massachusetts Academy of Math and Science

**Worcester, MA**

Ken Stafford, Brad Miller, Kevin Bobrowski, Alexander Hecht, Aaron Holroyd, Ryan Meador, Colleen Shaver, Paul Ventimiglia, Chris Werner, Stephanie Fuller, Danielle Jacobson, Steve Kaneb, Josh Mathews, Ciarán Murphy, Andrew Nehring, Dan Praetorius, Joshua Robins, Colin Roddy, David Tyler

### Team 237: Siemon Company, Plasti-Coat Corp., and Watertown High School

**Watertown, CT**

John Boucher, Cynthia Boucher, Jill Shaw, Jack Shaw, Gary Santoro, Sean Conway, Kurt Eckert, Elgin Clock, Sara Milton, Paul Hoffman, Everit Condit, Catherine Wilbur, Darrel Wilbur, Raymond Hebert Sr., Raymond Hebert Jr., Mark Henion, Scott Shaw, Kevin Shaw, Edward Boucher, Nicole Mikush, Elizabeth Beaudry, Brian Mendicino, Julie Henion, Michael Zoller, Michael Sorrenti, Michael Hoffman, Joseph Guerrera, Michael Lemay, Michael Preato, Matthew Voghel, Nicholas Fisher, Beth Sorrenti, Randy Sorrenti, Melanie Milton, Chlora Beaudry, Sue Ledell

### Team 294: Northrop Grumman, Mira Costa High School, and Redondo Union High School

**Redondo Beach, CA**

Eric Chumbley, Dale Hall, Bob Hosken, Pat Hosken, Bill Kunz, Ron Lee, Donald McKinzie, Mike McVey, LeRoy Nelson, Ricardo Salgado, Ken Sterk, Scott Tupper, Andrea Wagner, Rick Wagner, Kenny Agusta, Taylor Borel, Daniel Brim, Bryan Campbell, Andrew Cole, Matt Graham, Adam Heard, David Litwak, Stephen Ortiz, Greg Robinson, Alex Roth, David Tsao, Ryan Tupper, Thomas Wilson, Stefan Brown, Mike Treagar, Jade Karrillo

**Team 384: GE Volunteers, Qimonda of Richmond, ShowBest Fixture Corp., Specialty's Our Name, ChemTreat, Sams Club, CAPER, ITT Technical Institute, Henrico Co. Education Foundation, and Tucker High School**

**Richmond, VA**

Marshall Turner, Julie Norris, Carol Cotton, Leo Meire, Jeff Atkinson, Bob Benway, Terry Daniels, Greg Cotton, Jim Schubert, Chris Bandsfield, Janita Branson, Mike Terlease, Andrew Smith, Gaylan Nickey, Gigi Chan, Grayson Atkinson, Jackie Murrell, James McCarson, Jay Gowda, Jiten Narang, Jordan Lowell, Josh Griff, Justin Brown, Kien Tran, Krutica Kotval, Mae Cinco, Mary Jackson, Matt Lodge, Mitchell Cisney, Nadean Cubero, Nathan Moy, Peter Chen, Rameez Khimani, Rashidul Hason, Rick Lewis, Scott King, Sheeraja Rajagolpalan, Taey Duong, Thu-Thao, Trinh Tat, Will Gathright, Mike Milton, Larry Trelease, Driver Branson, Gwen Miller, Reggie Davenport

## Rockwell Automation Innovation in Control Award

**Team 111: Motorola, Rolling Meadows High School, and Wheeling High School**

**Schaumburg, IL**

Raul Olivera, Dan Posacki, Mike Soukup, Kristin Schultz, Dave Flowerday, Dave Scheck, Nathan Troup, Al Skierkiewicz, Surendra Devarashetty, Julie Atkins, Mark Rooney, George Graham, Miten Champaneri, Dan Green, Lisa Dohn, Erik Edhlund, Greg LaPrest, Vlad Voskoboynikov, Kirstie Cannon, Conor Delaney, Phillip Rybarczyk, Julio Salinas, Derek Solt, Tim Syoen, Tiffany Gach, Josh Koci, Rebecca Leung, Karl Swanson, Kate Thompson, Kyle Anderso, Michael Anderson, Brian Allred, Craig Babiarz, Courtney Balce, Robert Becker, Michael Borelli, Tomas Borosko, Mathias Brown, Jonathan Diores, Joseph Doose, Chris Fathke, Mark Faust, Aaron Glover, Daniel Hauswald, Kelly Hayden, Ryan Klaproth, Thomas Koehler, Brian Korves, Brian Kuczynski, John Kuehne, Mark Lenski, Jacob Mandozzi, Christopher Moss, Jason Park, Teagan Russell, Joe Rybarczyk, Alex Sabatka, Paul Simon, Sean Sy, Andy Tatkowski, Anthony Thompson, Kelli Vanantwerp, Brian Werling, Gregory Wilk, Adam Wojcik, Robert Zelm, Grant Farrand, Christina Olivera, Rich Olivera, Matt Simms, Bruce Johnson, Pam Balcer, Mark Faust, Brian Friberg, Wes Nurczyk, Phil Rybarczyk, Sharon Rybarczyk, Liza Sabatka, Dottie Skierkiewicz, Dana Solt, Tom Thompson, Frank Toussaint, Tim Werling, Jeff Bott, Patrick Fonsino, Mike Geist, Jeff Jerdee, Mark Koch, Jeff Vogt

**Team 225: School District of the City of York, NASA GSFC, Harley-Davidson of York, Legg-Mason Funds, Siemens Building Technologies, Cabin Fever Expo, William Penn High School, and William F. Goodling Regional Advanced Skills Center**

**York, PA**

Nam-Phuong, Marc, Won-Chik, LaToya, Nichelle, Theodore, Chris, Richard, Ian, Leah, Montez, Daniella, Hoai-Quoc, Sean, Brandon, Jesse, Derrick, Tabitha, Matt, Jeremiah, Jeremy, John, Aaron, Lizze, Lisa Tarman, St. Clair Simmons, Brandon May, Lisa Keys, Dien Buu, John Parker, Ron Karpinski, Tim Danner, Al Hoover, Stew Fink

**Team 418: LASA Robotics Association, National Instruments, and Liberal Arts and Science Academy of Austin**

**Austin, TX**

Jane Young, John Young, Danny Diaz, Randy Baden, Natalie Bixler, Ian Bonner, Judy DeWitt, Jennifer Gibbons, Eric Giesberg, Lewis Henrich, Tracey Janus, Harrison Key, Karthish Manthiram, Henry Mareck, Ryan Newton, Clint Olmos, David Riffey, Gerry Salinas, Rahul Sinha, Rohit Sinha, Michael Singerman, Allen Smith, Billy Sweetman, Cathy Thiele, Price Vetter, Mitchel Wilkinson

**Team 494: DaimlerChrysler, General Motors, Textron Fastening Systems, and Goodrich High School**

**Goodrich, MI**

Thomas Baker, Nick Born, Alec Bretzloff, Clint Brinker, Tim Burr, Jameson Carbary, David Dew, Billy Dixon, Zack Downer, Ian Gabrielson, Carly Goerke, Will Hubbard, Noah Lane, Justin Miller, Jack Nowakowski, Ryan Starski, Jimmy Tyner, Nathan Wallace, Eva Yokosawa, Sofia Yokosawa, Matthew Zagone. Kay Baker, Ed & Kelly Bretzloff, Jim Carbury, Clint Densham, Bill Dixon, Jack Frost, Tom & Nancy Gabrielson, Patrick & Patricia Major, Bill Maxwell, Art & Liz Nowakowski, Jay & Gaye TenBrink, Greg Thurk, Dave Wallace, Gene & Mona Williams, Eric Woolley, Ken & Tina Yokosawa, Nancy Zagone

**Team 1114: General Motors—St. Catharine's Powertrain, and Governor Simcoe Secondary School**

**St. Catharines, ON Canada**

T.J. Andrade, Jeff Beckett, Zac Corfield, Lindsay Davies, Peter Diakow, Melissa Doornekamp, Stephen Doornekamp, Ian Froud, Jesse Graham, Julia Greco, Mark Jen, Corey LeBlanc, Eric LePalud, Chris Lyddiatt, Stuart MacGregor, David Mikolajewski, Hannah Mitchell, Zak Mason, Kate Mosley, Loretta Peterson, Alex Post, Brandon Pruniak, Jessa Pruniak, Nik Unger, Luke Visser, Jordan Wakulich, Mike Williams, Bob Allan, Geoff Allan, Derek Bessette, Emerald Church, Michael DiRamio, John Froud, Steve Hauck, Tyler Holtzman, Karthik Kanagasabapathy, Esther King, Ian Mackenzie, Kate MacNamara, Don Mason, Greg Phillips, Joanne Pruniak, Mike Pruniak, Steve Rourke, Brent Selvig, Gary Unger, Matt Vint, Todd Willick

**Team 1629: NASA, Beitzel Corporation, and Garrett County Public Schools**

**McHenry, MD**

Titus Beitzel, Rick Dolan, Larry Friend, Josh Hinebaugh, Rich Lewis, Phil Malone, Larry Mullenax, Eric Perfetti, Chuck Trautwein, Philip Adams, Devynn Brant, Jed Crawford, Sarah Coddington, Mitch Hall, J.P. Law, James Mullenax, Emily Roser, Zachary Trautwein, Sarah Storck, Stephanie Lee, Chris Wood, Erik Wood

## Motorola Quality Award

**Team 16: Mountain Home High School**

**Mountain Home, AR**

Amber Schulz, Josh Shelly, Jordan Wilhite, Bryce Caldwell, Eric Normandy, Kayleigh Brown, JT Easley, Nicole Moore, Galen Doyel, Ryan Croom, Jaime Ott, Jake Eickman, Kolin Paulk, Michael Griffin, Zach Bogart, Sarah Bentley, Heath Hughes, Chris Marts, Chris Parks, Michelle Lunday, Kyle Markowski, Clayton Brinza, Kevin Adams, Derrick Williamson, Chris Knight, Mario Hernandez, Andrew Tuberville, Steve Hatch, Ray Faulkner, Michael Lunday, Jackie Meissner, Brandon Padgett, Dr. Cliff Croom, Tim Maynard, David Thompson, John Novak, Mike DeMass, Tom Goates, Andy Marts, Mike Osmundson, Dr. Stephen Vester

**Team 141: JR Automation Technologies, Inc. and West Ottawa High School**

**Holland, MI**

Norma Lamotte, Pat Lamotte, Pete Mokris, Susan Smith, Selden Smith, Betty DeLaRosa, Bob Schafer, Cheryl Walters, Gary Walters, Kelly Walters, Kelle Prescott, Evan Sharp, David Barr, Dan Mouw, Beulah Telman, Justin Telman, David Smeenge, Doug Papay, Carrrie Papay, Dee Kozumplik, John Kozumplik, Joe Straznac, Steven Gillette, Kyle Keusel, Phi Dao, Drew DeWit, David Smith, Max Mills, Dennis Pardy, Vaughn Schrotenboer, Mark Sprietzer, Debbie Zwiers, Trent Jackson, Matt Barr, Cory Walters, Anthony Corder, Savannah Weaver, Liz Seeber, David Michael Smeenge, Brynn Sharp, Claire Kinne, Nicole Schipper, Ryan Bozio, Brandon Dams, Cassie Telman, Bonnie Lamotte

**Team 191: Xerox Corporation and Joseph C. Wilson Magnet High School**

**Rochester, NY**

Ronald Dukes, Peg Foos, Joanne Mannix, Willie Robinson, Carlos Terrero, Ellery Wong, Rosaria D'Aiuto, Datwan Dixon, Andre Martin, Jeffrey Huspen, Duane Jennings, Carl Lewis, Murray Meetze, Edward Patterson, Linda Wallace, Jamon T. Williams, Rebecca Wong, Samantha Bommelje, Rebecca Bostock, Kelsey Carlson, Kevin Chaba, Jonathan Clark, Matthew Clark, Rosanna Doyle, Raina Edwards, Meredith Emerson, Axel Engle, Patricia Giron, Kyle Gordon, Jamaal Johnson, Kenneth Johnson, Henry Jones, Cashara Kelley, Alexandra Leonard, Jay Lopez, Jovany Martinez, Timothy McCrossen, Mark Rand, Karla Molinero, Marine Mukashambo, Ngoc Nguyen, Sarah Rebholz, Peter Reineger, Herminio Rivera, Charlene Sanchez, Katie Sarubbi, Casey Sukahavong

**Team 207: Walt Disney Imagineering and Centinela Valley Union High School District**

**Hawthorne, CA**

Richardo Acosta, Josue Arias, Jose Alejo, Nicholas Atain, Bianca Campos, Adrian Castaneda, Luisiana Alfaro, Jose Ceja, Xenia Davis, Sean Drury, Chrystian Bautista, Joeseph Egusquiza, Luis Figueroa, Nary Guardado, Kimberly Crespin, Brette Hanna, Warren Meyer, Christian Orozco, Rogelio Paz, Jesus Marroquin, Liliana Sandoval, William Valverde, Miguel Varela, Myron DeCloud, Christina Wells, Emely Morales, Salvador Sandoval, David Torres, Edward Alfaro, Steve Ferron, Jason Takamoto, Leanne Haarbauer, Caline Khavarani, Chuck Knight, Lucas Pacheco, John Depillo, Kyle Depillo, Don Metz, Dee Murr, Michael Gordon, Ed Memeth

**Team 322: General Motors Powertrain, Landaal Packaging, University of Michigan—Flint, Flint Community Schools, and GEAR-UP**

**Flint, MI**

Dan Clemons, Kelly Buetow, Dave Buetow, Russ Perkins, Gary Mueller, Marcus Hunter, Rashawn Hunter, Brandon Barnes, Willy Strong, Doug Keith

**Team 1024: Aircom Manufacturing, Beckman Coulter, Rolls-Royce Corporation, and Bernard K. McKenzie Career Center**

**Indianapolis, IN**

Erik Buland, Libby Glass, Matt Swanson, Michael Long, Anne Schneider, Matt Miller, Tommy Francis, Jennifer Maberto, Nick Bandy, David Badger, Jake Gass, Claire Larew, Marc Rhodes, Bill Townsend, Talyor Chenoweth, Chad Heck, Adam Suchko, Alex Suchko, Jeffrey Smith, Allison Metzger, Grant Lin, Jason Zielke, Zach Farr, Andrew DeFeo, Melvin Harris, Bradley Thomason, Weston Hoskins, Michael Littell, Michael Schevitz, Ben Nelson, Chris Boyd, Josh Allanson, Tom Gentry, Zach Herman, Kristin Koch, Jared Niederhauser, Brett Wolfsie, Gary Valler, Mike VanVertloo, Bruce Long

## Xerox Creativity Award

**Team 86: JEA, Johnson & Johnson, VISTAKON, & Stanton College Preparatory School**

**Jacksonville, FL**

Lisa Lovelace, Todd Lovelace, Dave Vogel, Glenn Williams, Bill Kearson, Jacob Plicque, Aaron Paulson, Clifford Newkirk, Trevor Ollar, Alex Brito, Wil Brito, Jimmy Tomola, Zach Michael, Cory Hester, Lyle Josey, Thomas Manning, Brandon Ballew, Matt Goldstein, Scott Manning, Jim Warren, Joachim Hartje, Billy Herod, Catherine Daas, Brad Paulson, Brian Schubert, Jean Schubert, Marcus Snyder, Mark McCombs, Kyle Bivens, Harold Hou, Jeff Edwards, Chris Edwards, Linda Edwards

**Team 293: Bristol-Myers Squibb, MEI, and Hopewell Valley Central High School**

**Pennington, NJ**

James Andrews, Glen Babecki, Gordon Brooks, John Delaney, Gale Downey, Joe Knotts, Sky Morehouse, Sam Ornstein, Joe Nolfo, Harry Peles, Ed Pitrillo, Joe Sinniger, Mike Thibeault, Mary Yeomans, Manisika Agaskar, Zac Andrews, Christopher Babecki, Michael Bolan, Danny Brown, Cole Gargione, Max Gilbert, Ardres Gonzolez, Bryan Hatch, Simon Healey, Owen Healy, David Hill, Gus Hubner, Ben Jarrett, Christina Jaworsky, Han-wei Kautzer, Noah Kershaw, Max Konig, Peter Krasucki, Sophie Krasucki, Robert LaPosta, Michael Lewis, Lisa Liu, Amalie McKee, Alex Melman, Bobby Miller, Bichael Ornstein, Joseph Patuick, Archana Rachakonda, Ellen Sherwood, Martin Sherwood, Scot Stilson, Megan Templon, Ben Van Selas, Alex Yang

**Team 330: J&F Machine, NASA-JPL, NASA-Goddard, Raytheon, and Hope Chapel Academy**

**Hermosa Beach, CA**

Larry Couch, David Drennon, Matt Driggs, Steve Eccles, Tom Freitag, Bill Hendry, Chris Husmann, Michael Palmerino, Ben Roberts, Ric Roberts, Greg Ross, Joe Ross, Rob Steele, Zach Steele, Jacob Trimper, Rick Varnum, Sam Couch, David Freitag, Eric Husmann, Elijah Mounts, Shane Palmerino, Courtney Roberts, Curtis Rose, Andy Ross, Caitlin Steele, Ethan Wilson, Penny Ross

**Team 694: Credit Suisse, D. E. Shaw & Co., Yvette & Larry Gralla, Con Edison, Time, Inc., The Wallace Foundation, Cox & Company, Inc., Stuyvesant High School Alumni Association, Parents Association, and Stuyvesant High School**

**New York, NY**

Rafael Colón, Paul Desiderio, Ian Ferguson, Tom Ferguson, Mel Hauptman, Steve Hilton, Colin Holgate, Jesse Hong, Wendy Keyes, Catherine Kunicki, Ron Kunicki, Abigail Laufer, Justin Lee, James Lonardo, Joe Ricci, Andy Woo, Nancy Yabroudi, Anat Zloof, Tal Akabas, Daniel Alzugaray, Katie Banks, Nathan Bixler, Joe Blay, Josef Broder, Joshua Budofsky, Steven Cao, Samuel Crisanto, Jena Cutie, Polina Danilouk, Paul Desiderio, Allan Dong, Joshua Hecht, Tala Huhe, Ethan Illfelder, Edward Kaplan, Jenna Kefeli, Sarah Ketani, Nathan Keyes, Theodora Kunicki, Andrew LaBunka, Steven Lam, Benjamin Lee, Yi Li, Peter Liu, Victor Liu, Joanna Ma, Andrew Mandelbaum, Jonathan Meed, Kimberly Milner, Manav Nanda, Sakif Noor, Jason Rassi, Diana Sandy, Olga Shishkov, Amy Suen, Tiffany Tsai, William Twomey, Sho Uemura, Daryl Vulis, Jay Walker, Jesse Weinman, Harrison Wong, Sami Yabroudi, Dishen Yang, Flynn Zaiger, Danny Zhu, Maya Zloof, Yon Zloof

**Team 1319: AdvanceSC, Laughlin Racing Products, Sealed Air Corp. Cryovac, AssetPoint, National Electrical, Greenville County Schools, and Mauldin High School**

**Mauldin, SC**

Erich Zende, Robert Zende, Chuck Zende, Nancy Zende, Rick Wilson, Glenn Killinger, Hugh Rambo, Jeff Corbett, Patrick Robert, Anna Robert, Leesa Brotherton, Crystal Dickerson, Catherine Zende, Elliot Dickerson, Travis Suttles, Mitch Neiling, Alex Slessman, Matt Wilson, Mason Walker, Trace Guy, Zack Ayres, Stephen Corbett, Alex Kelly, Adam Sherrick, Ethan Suttles, Ryan Upham, Ian Harris, Jon Gawrych

**Team 1510: NASA, Beaverton Education Foundation, Intel, Tektronix, PGE, Portland Community College, and Westview Robotics Team**

**Beaverton, OR**

Caio Tenca, Anjali Menon, Andrew Fagin, Andy Goetz, Andy Hsiao, Arron Ho, Ben Stolt, Brent Pugh, Christopher Vergara, CJ Hyde, David Kresta, Jasmine Sears, Keegan Edwards, Nolan Check, Prachi Pai, Rajesh Gunaji, Rohit Saxena, Steven Garcia, Tanmay Bhardwaj, Thomas Joseph, Yujie Zeng, Williane Tenca, Alexandre Tenca, Mark Garcia, Alex Knots, Robert Hendel, David Locke, Gladys Chin, Jean Tenca, Patrick Stolt, Tom Edwards, Pai Janardan, Patrick Kraft, Matthew Scott, Michael Flaman

## ⇄ Photographers and Illustrators

Each profiled team supplied the photographs and illustrations that appear in this book.

Additional photograph and illustration credits:
Adriana M. Groisman, FIRST
Joe Menassa, Joe Menassa Photography
Dave Lavery, NASA
Innovation First, Incorporated